Using R for Statistics

Sarah Stowell

Apress®

Using R for Statistics

Copyright © 2014 by Sarah Stowell

This work is subject to copyright. All rights are reserved by the Publisher, whether the whole or part of the material is concerned, specifically the rights of translation, reprinting, reuse of illustrations, recitation, broadcasting, reproduction on microfilms or in any other physical way, and transmission or information storage and retrieval, electronic adaptation, computer software, or by similar or dissimilar methodology now known or hereafter developed. Exempted from this legal reservation are brief excerpts in connection with reviews or scholarly analysis or material supplied specifically for the purpose of being entered and executed on a computer system, for exclusive use by the purchaser of the work. Duplication of this publication or parts thereof is permitted only under the provisions of the Copyright Law of the Publisher's location, in its current version, and permission for use must always be obtained from Springer. Permissions for use may be obtained through RightsLink at the Copyright Clearance Center. Violations are liable to prosecution under the respective Copyright Law.

ISBN-13 (pbk): 978-1-4842-0140-4

ISBN-13 (electronic): 978-1-4842-0139-8

Trademarked names, logos, and images may appear in this book. Rather than use a trademark symbol with every occurrence of a trademarked name, logo, or image we use the names, logos, and images only in an editorial fashion and to the benefit of the trademark owner, with no intention of infringement of the trademark.

The use in this publication of trade names, trademarks, service marks, and similar terms, even if they are not identified as such, is not to be taken as an expression of opinion as to whether or not they are subject to proprietary rights.

While the advice and information in this book are believed to be true and accurate at the date of publication, neither the authors nor the editors nor the publisher can accept any legal responsibility for any errors or omissions that may be made. The publisher makes no warranty, express or implied, with respect to the material contained herein.

Publisher: Heinz Weinheimer
Lead Editor: Steve Anglin
Development Editor: Matthew Moodie and Chris Nelson
Technical Reviewers: Myron Hlynka, Wen Sui Liu, and Larry Pace
Editorial Board: Steve Anglin, Mark Beckner, Ewan Buckingham, Gary Cornell, Louise Corrigan, Jim DeWolf, Jonathan Gennick, Jonathan Hassell, Robert Hutchinson, Michelle Lowman, James Markham, Matthew Moodie, Jeff Olson, Jeffrey Pepper, Douglas Pundick, Ben Renow-Clarke, Dominic Shakeshaft, Gwenan Spearing, Matt Wade, Steve Weiss
Coordinating Editor: Anamika Panchoo
Copy Editor: Laura Lawrie
Compositor: SPi Global
Indexer: SPi Global
Artist: SPi Global
Cover Designer: Anna Ishchenko

Distributed to the book trade worldwide by Springer Science+Business Media New York, 233 Spring Street, 6th Floor, New York, NY 10013. Phone 1-800-SPRINGER, fax (201) 348-4505, e-mail orders-ny@springer-sbm.com, or visit www.springeronline.com. Apress Media, LLC is a California LLC and the sole member (owner) is Springer Science + Business Media Finance Inc (SSBM Finance Inc). SSBM Finance Inc is a Delaware corporation.

For information on translations, please e-mail rights@apress.com, or visit www.apress.com.

Apress and friends of ED books may be purchased in bulk for academic, corporate, or promotional use. eBook versions and licenses are also available for most titles. For more information, reference our Special Bulk Sales–eBook Licensing web page at www.apress.com/bulk-sales.

Any source code or other supplementary material referenced by the author in this text is available to readers at www.apress.com/9781484201404. For detailed information about how to locate your book's source code, go to www.apress.com/source-code/.

Contents at a Glance

Contents

About the Author

Sarah Stowell is a contract statistician based in the UK, who has worked with Mitsubishi Pharma Europe, MDSL International, and GlaxoSmithKline previously. She holds a Master of Science degree in Statistics.

About the Technical Reviewer

Dr. Larry Pace is a statistics author and educator as well as a consultant. He lives in the upstate area of South Carolina in the town of Anderson. He is a professor of statistics, mathematics, psychology, management, and leadership. He has programmed in a variety of languages and scripting languages including R, Visual Basic, JavaScript, C##, PHP, APL, and, in a long-ago world, Fortran IV. He writes books and tutorials on statistics, computers, and technology. He has also published many academic papers, and made dozens of presentations and lectures. He has consulted with Compaq Computers, AT&T, Xerox Corporation, the U.S. Navy, and International Paper. He has taught at Keiser University, Argosy University, Capella University, Ashford University, Anderson University (where he was the chair of the behavioral sciences department), Clemson University, Louisiana Tech University, LSU in Shreveport, the University of Tennessee, Cornell University, Rochester Institute of Technology, Rensselaer Polytechnic Institute, and the University of Georgia.

Acknowledgments

First, I would like to thank the Apress team, in particular: Lead Editor Steve Anglin, for getting me on board and giving me the chance to work with Apress; Coordinating Editors Anamika Panchoo and Mark Powers for keeping me on track; Development Editor Chris Nelson for teaching me a lot about writing; Technical Editor Larry Pace for making many valuable suggestions to improve the quality of the book; and the many others whom I have not met but I can see have done a great job helping to create the finished product.

I would also like to thank to Andrés Barnett, James Sedgwick, and Therese Stukel for providing data for the examples, and my husband Timothy Baldock and friends Jemma-Kay Johnstone, Christopher Gilmour, Nina Farrell, Chris Brown, Artur Kyral, and Eddie Chung, who have all helped with the project in its early stages.

Introduction

Welcome to *Using R for Statistics*. This book was written for anyone who wants to use R to analyze data and create statistical plots. It is suitable for those with little or no experience with R, and aims to get you up and running quickly without having to learn all the details of programming.

About R

R is a statistical analysis and graphics environment and also a programming language. It is command-driven and very similar to the commercially produced S-Plus® software. R is known for its professional-looking graphics, which allow complete customization.

R is open-source software and free to install under the GNU general public license. It is written and maintained by a group of volunteers known as the R core team.

The base software is supplemented by over 5,000 add-on packages developed by R users all over the world, many of whom belong to the academic community. These packages cover a broad range of statistical techniques including some of the most recently developed and niche purpose. Anyone can contribute add-on packages, which are checked for quality before they are added to the collection.

At the time of writing, the current version of R is 3.1.0.

What You Will Learn

This book is designed to give straightforward, practical guidance for performing popular statistical methods in R. The programming aspect of R is explored only briefly.

After reading this book you will be able to:

- navigate the R system

- enter and import data

- manipulate datasets

- calculate summary statistics

- create statistical plots and customize their appearance

- perform hypothesis tests such as the t-test and analysis of variance

- build regression models

- access additional functionality with the use of add-on packages

- create your own functions

Knowledge Assumed

Although this book does include some reminders about statistics methods and examples demonstrating their use, it is not intended to teach statistics. Therefore, you will require some previous knowledge. You should be able to select the most appropriate statistical method for your purpose and interpret the results. You should also be familiar with common statistical terms and concepts. If you are unsure about any of the methods that you are using, I recommend that you use this book in conjunction with a more detailed book on statistics.

No prior knowledge of R or of programming is assumed, making this book ideal if you are more accustomed to working with point-and-click style packages. Only general computer skills and a familiarity with your operating system are required.

Conventions Used in This Book

This book uses the following typographical conventions:

- Fixed width font is used to distinguish all R commands and output from the main text.

- Normal fixed width font is used for built-in R function names, argument names, syntax, specific dataset and variable names, and any other parts of the commands that can be copied verbatim.

- Slanted fixed width font is used for generic dataset and variable names and any other parts of the commands that should be replaced with the user's own values.

- Often it has not been possible to fit a whole command into the width of the page. In these cases, the command is continued on the following line and indented. Where you see this, the command should still be entered into the console on a single line.

- Text boxes, which are separate from the main text, contain reminders of statistical theory or methods.

- Practical examples are presented in separate numbered sections.

Datasets Used in This Book

A large number of example datasets are included with R, and these are available to use as soon as you open the software. This book makes use of several of these datasets for demonstration purposes.

There are also a number of additional datasets used throughout the book, details of which are given in the Appendix C. They are available to download at www.apress.com/9781484201404.

Contact the Author

If you have any suggestions or feedback, I would love to hear from you. You can email me at s.stowell@instantr.com.

CHAPTER 1

■ ■ ■

R Fundamentals

R is a statistical analysis and graphics environment that is comparable in scope to the SAS, SPSS, Stata, and S-Plus packages. The basic installation includes all of the most commonly used statistical techniques such as univariate analysis, categorical data analysis, hypothesis tests, generalized linear models, multivariate analysis, and time-series analysis. It also has excellent facilities for producing statistical graphics. Anything not included in the basic installation is usually covered by one of the thousands of add-on packages available.

Because R is command-driven, it can take a little longer to master than point-and-click style software. However, the reward for your effort is the greater flexibility of the software and access to the most newly developed methods.

To get you started, this chapter introduces the R system. You will:

- download and install R

- become familiar with the interface

- start giving commands

- learn about the different types of R files

- become familiar with all of the important technical terms that will be used throughout the book

If you are new to R, I recommend that you read the entire chapter, as it will give you a solid foundation on which to build.

Downloading and Installing R

The R software is freely available from the R website. Windows® and Mac® users should follow the instructions below to download the installation file:

1. Go to the R project website at www.r-project.org.

2. Follow the link to CRAN (on the left-hand side).

3. You will be taken to a list of sites that host the R installation files (mirror sites). Select a site close to your location.

4. Select your operating system. There are installation files available for the Windows, Mac, and Linux® operating systems.

5. If downloading R for Windows, you will be asked to select from the base or contrib distributions. Select the base distribution.

6. Follow the link to download the R installation file and save the file to a suitable location on your machine.

To install R for the Windows and Mac OS environments, open the installation file and follow the instructions given by the setup wizard. You will be given the option of customizing the installation, but if you are new to R, I recommend that you use the standard installation settings. If you are installing R on a networked computer, you may need to contact your system administrator to obtain permission before performing the installation.

For Linux users, the simplest way to install R is via the package manager. You can find R by searching for "r-base-core." Detailed installation instructions are available in the same location as the installation files.

If you have the required technical knowledge, then you can also compile the software from the source code. An in-depth guide can be found at `www.stats.bris.ac.uk/R/doc/manuals/R-admin.pdf`.

Getting Orientated

Once you have installed the software and opened it for the first time, you will see the R interface as shown in Figure 1-1.

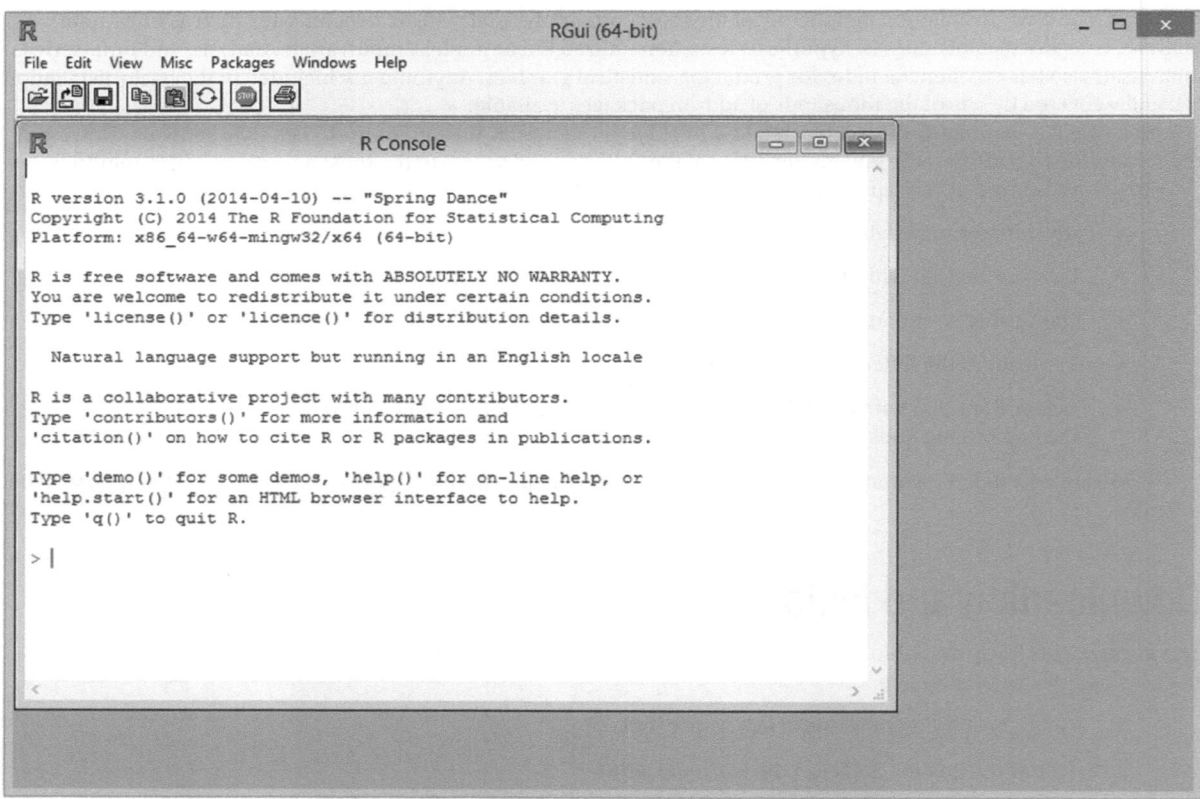

Figure 1-1. *The R interface*

There are several drop-down menus and buttons, but unlike in point-and-click style statistical packages, you will only use these for supporting activities such as opening and saving R files, setting preferences, and loading add-on packages. You will perform all of the main tasks (such as importing data, performing statistical analysis, and creating graphs) by giving R typed *commands*.

The R Console window is where you will type your commands. It is also where the output and any error messages are displayed. Later you will use other windows such as the data editor, script editor, and graphics device.

The R Console and Command Prompt

Now turn your attention to the R console window. Every time you start R, some text relating to copyright and other issues appears in the console window, as shown in Figure 1-1. If you find the text in the console difficult to read, you can adjust it by selecting GUI Preferences from the Edit menu. This opens a dialog box that allows you to change the size and font of the console text, as well as other options.

Below all of the text that appears in the console at startup you will see the *command prompt*, which is colored red and looks like this:

>

The command prompt tells you that R is ready to receive your command.

Try typing the following command at the prompt and pressing Enter:

> 8-2

R responds by giving the following output in the next line of the console:

[1] 6
>

The [1] tells you which component of the output you are looking at, which is not of much interest at this stage as the output has only one component. This is followed by the result of the calculation, which is 6. Notice that all output is shown in blue, to distinguish it from your commands.

The output is followed by another prompt > to tell you that it has finished processing your command and is ready for the next one. If you don't see a command prompt after entering a command, it may be because the command you have given is not complete. Try entering the following incomplete command at the command prompt:

> 8-

R responds with a plus sign:

+

If you see the plus sign, it means you need to type the remainder of the command and press Enter. Alternatively, you can press the Esc key to cancel the command and return to the command prompt.

Another time that you would not see the command prompt is when R is still working on the task. Usually this time is negligible, but there may be some waiting time for more complex tasks or those involving large datasets. If a command takes much longer than expected to complete, you can cancel it with the Esc key.

From here onward, the command prompt will be omitted when showing output.

Table 1-1 shows the symbols used to represent the basic arithmetic operations.

Table 1-1. *Arithmetic Operators*

Operation	Symbol
Addition	+
Subtraction	-
Multiplication	*
Division	/
Exponentiation	^

If a command is composed of several arithmetic operators, they are evaluated in the usual order of precedence, that is, first the exponentiation (power) symbol, followed by division, then multiplication, and finally addition and subtraction. You can also add parentheses to control precedence if required. For example, the command:

```
> 3^2+6/3+2
```

gives the result:

```
[1] 13
```

while the command:

```
> (3^2+6)/(3+2)
```

gives the result:

```
[1] 3
```

If you want to repeat a command, you can use the up and down arrow keys on your keyboard to scroll through previously entered commands. You will be able to edit the command before pressing Enter. This means that you don't have to retype a whole command just to correct a minor mistake, which you will find useful as you begin to use longer and more complex commands.

Functions

In order to do anything more than basic arithmetic calculations, you will need to use *functions*. A function is a set of commands that have been given a name and together perform a specific task producing some kind of output. Usually a function also requires some kind of data as input.

R has many built-in functions for performing a variety of tasks from simple things like rounding numbers, to importing files and performing complex statistical analysis. You will make use of these throughout this book. You can also create your own functions, which is covered briefly in Chapter 12.

Whenever you use a function, you will type the function name followed by round brackets. Any input required by the function is placed between the brackets.

An example of a function that does not require any input is the date function, which gives the current date and time from your computer's clock.

```
> date()
```

```
[1] "Thu Apr 10 20:59:26 2014"
```

An example of a simple function that requires input is the round function, which rounds numbers. The input required is the number you want to round. A single piece of input is known as an *argument*.

```
> round(3.141593)
```

```
[1] 3
```

As you can see, the round function rounds a given number to the nearest whole number, but you can also use it to round a number to a different level of accuracy. The command below rounds the same number to two decimal places:

```
> round(3.141593, digits=2)
```

```
[1] 3.14
```

We were able to change the behavior of the round function by adding an additional argument giving the number of decimal places required. When you provide more than one argument to a function, they must be separated with commas. Each argument has a name. In this case, the argument giving the number of decimal places is called digits. Often you don't need to give the names of the arguments, because R is able to identify them by their values and the order in which they are arranged. So for the round function, the following command is also acceptable:

```
> round(3.141593, 2)
```

Some arguments are optional and some must be provided for the function to work. For the round function, the number to be rounded (in this example 3.141593) is a required argument and the function won't work without it. The digits argument is optional. If you don't supply it, R assumes a default value of zero.

For every function included with R, there is a help file that you can view by entering the command:

```
> help(functionname)
```

The help file gives details of all of the arguments for the function, whether they are required or optional and what their default values are.

Table 1-2 shows some useful mathematical functions.

Table 1-2. *Useful Mathematical Functions*

Purpose	Function
Exponential	exp
Natural logarithm	log
Log base 10	log10
Square root	sqrt
Cosine	cos
Sine	sin
Tangent	tan
Arc cosine	acos
Arc sine	asin
Arc tangent	atan
Round	round
Absolute value	abs
Factorial	factorial

Objects

In R, an *object* is some data that has been given a name and stored in the memory. The data could be anything from a single number to a whole table of data of mixed types.

Simple Objects

You can create objects with the *assignment operator*, which looks like this:

```
<-
```

For example, to create an object named height that holds the value 72.95 (a person's height in inches), use the command:

```
> height<-72.95
```

When creating new objects, you must choose an object name that:

- consists only of upper and lower case letters, numbers, underscores (_) and dots (.)
- begins with an upper- or lowercase letter or a dot (.)
- is not one of R's reserved words (enter help(reserved) to see a list of these)

R is case-sensitive, so height, HEIGHT, and Height are all distinct object names.

If you choose an object name that is already in use, you will overwrite the old object with the new one. R does not give any warning if you do this.

To view the contents of an object you have already created, enter the object name:

```
> height
```

```
[1] 72.95
```

Once you have created an object, you can use it in place of the information it contains. For example, as input to a function:

```
> log(height)
```

```
[1] 4.289774
```

As well as creating objects with specific values, you can save the output of a function or calculation directly to a new object. For example, the following command converts the value of the height object from inches to centimeters and saves the output to a new object called heightcm:

```
> heightcm<-round(height*2.54)
```

Notice that when you assign the output from a function or calculation to an object, R does not display the output. To see it, you must view the contents of the object by entering the object name.

To change the contents of an object, simply overwrite it with a new value:

```
> height<-69.45
```

Objects like these are called *numeric* objects because they contain numbers. You can also create other types of objects such as *character* objects, which contain a combination of any keyboard characters known as a *character string*. When creating a character object, enclose the character string in quotation marks:

```
> string1<-"Hello!"
```

You can use either double or single quotation marks to enclose a character string, as long they are both of the same type. To include quotation marks within a character string, place a backslash before the quotes (known as an *escape sequence*):

```
> string2<-"I said \"Hello!\""
```

So far we have only discussed simple objects that contain a single data value, but you can also create more complex types of objects. Two important types are *vectors* and *data frames*. A vector is an object that contains several data values of the same type. A data frame is an object that holds an entire dataset. Vectors and data frames are discussed in more detail in the following sections.

Vectors

A vector is an object that holds several data values of the same type arranged in a particular order. You can create vectors with a special function which is named c. For example, suppose that you have measured the temperature in degrees centigrade at five randomly selected locations and recorded the data as: 3, 3.76, -0.35, 1.2, -5. To save the data to a vector named temperatures, use the command:

```
> temperatures<-c(3, 3.76, -0.35, 1.2, -5)
```

You can view the contents of a vector by entering its name, as you would for any other object.

```
> temperatures
```

```
[1]   3.00   3.76 -0.35   1.20 -5.00
```

The number of values a vector holds is called its *length*. You can check the length of a vector with the length function:

```
> length(temperatures)
[1] 5
```

Each data value in the vector has a position within the vector, which you can refer to using square brackets. This is known as *bracket notation*. For example, you can view the third member of temperatures with the command:

```
> temperatures[3]
```

```
[1] -0.35
```

If you have a large vector (such that when displayed, the values of the vector fill several lines of the console window), the indices at the side tell you which member of the vector each line begins with. For example, the vector below contains twenty-seven values. The indices at the side show that the second line begins with the eleventh member and the third line begins with the twenty-first member. This helps you to determine the position of each value within the vector.

```
[1]    0.077  0.489  1.603  2.110  2.625  1.019  1.100  1.729  2.469 -0.125
[11]   1.931  0.155  0.572  1.160 -1.405  2.868  0.632 -1.714  2.615  0.714
[21]   0.979  1.768  1.429 -0.119  0.459  1.083 -0.270
```

If you give a vector as input to a function intended for use with a single number (such as the exp function), R applies the function to each member of the vector individually and gives another vector as output:

```
> exp(temperatures)
```

```
[1] 20.085536923 42.948425979   0.704688090   3.320116923   0.006737947
```

Some functions are designed specifically for use with vectors and use all members of the vector together to create a single value as output. An example is the mean function, which calculates the mean of all the values in the vector:

```
> mean(temperatures)
```

```
[1] 0.522
```

The mean function and other statistical summary functions are discussed in more detail in Chapter 5.

Like basic objects, vectors can hold different types of data values such as numbers or character strings. However, all members of the vector must be of the same type. If you attempt to create a vector containing both numbers and characters, R will convert any numeric values into characters. Character representations of numbers are treated as text and cannot be used in calculations.

Data Frames

A data frame is a type of object that is suitable for holding a dataset. A data frame is composed of several vectors of the same length, displayed vertically and arranged side by side. This forms a rectangular grid in which each column has a name and contains one vector. Although all of the values in one column of a data frame must be of the same type, different columns can hold different types of data (such as numbers or character strings). This makes them ideal for storing datasets, with each column holding a variable and each row an observation.

In Chapter 2, you will learn how to create new data frames to hold your own datasets. For now, there are some datasets included with R that you can experiment with. One of these is called Puromycin, which we will use here to demonstrate the idea of a data frame. You can view the contents of the Puromycin dataset in the same way as for any other object, by entering its name at the command prompt:

```
> Puromycin
```

R outputs the contents of the data frame:

	conc	rate	state
1	0.02	76	treated
2	0.02	47	treated
3	0.06	97	treated
4	0.06	107	treated
5	0.11	123	treated
6	0.11	139	treated
7	0.22	159	treated
8	0.22	152	treated
9	0.56	191	treated
10	0.56	201	treated
11	1.10	207	treated
12	1.10	200	treated
13	0.02	67	untreated
14	0.02	51	untreated
15	0.06	84	untreated
16	0.06	86	untreated
17	0.11	98	untreated
18	0.11	115	untreated
19	0.22	131	untreated
20	0.22	124	untreated
21	0.56	144	untreated
22	0.56	158	untreated
23	1.10	160	untreated

The dataset has three variables named conc, rate, and state, and it has 23 observations.

It is important to know how to refer to the different components of a data frame. To refer to a particular variable within a dataset by name, use the dollar symbol ($):

```
> Puromycin$rate
```

```
 [1]  76  47  97 107 123 139 159 152 191 201 207 200  67  51  84  86  98 115 131
[20] 124 144 158 160
```

This is useful because it allows you to apply functions to the variable, for example, to calculate the mean value of the rate variable:

```
> mean(Puromycin$rate)
```

```
[1] 126.8261
```

As well as selecting variables by name with the dollar symbol, you can refer to sections of the data frame using bracket notation. Bracket notation can be thought of as a coordinate system for the data frame. You provide the row number and column number between square brackets:

```
> dataset[r,c]
```

For example, to select the value in the sixth row of the second column of the Puromycin dataset using bracket notation, use the command:

```
> Puromycin[6,2]
```

```
[1] 139
```

You can select a whole row by leaving the column number blank. For example to select the sixth row of the Puromycin dataset:

```
> Puromycin[6,]
```

```
  conc  rate    state
6 0.11   139  treated
```

Similarly to select a whole column, leave the row number blank. For example, to select the second column of the Puromycin dataset:

```
> Puromycin[,2]
```

```
 [1]  76  47  97 107 123 139 159 152 191 201 207 200  67  51  84  86  98 115 131
[20] 124 144 158 160
```

When selecting whole columns, you can also leave out the comma entirely and just give the column number.

```
> Puromycin[2]
```

	rate
1	76
2	47
3	97
4	107
5	123
6	139
7	159
8	152
9	191
10	201
11	207
12	200
13	67
14	51
15	84
16	86
17	98
18	115
19	131
20	124
21	144
22	158
23	160

Notice that the command Puromycin[2] produces a data frame with one column, while the command Puromycin[,2] produces a vector.

You can use the minus sign to exclude a part of the data frame instead of selecting it. For example, to exclude the first column:

```
> Puromycin[-1]
```

You can use the colon (:) to select a range of rows or columns. For example, to select row numbers six to ten:

```
> Puromycin[6:10,]
```

	conc	rate	state
6	0.11	139	treated
7	0.22	159	treated
8	0.22	152	treated
9	0.56	191	treated
10	0.56	201	treated

To select nonconsecutive rows or columns, use the c function inside the brackets. For example, to select columns one and three:

```
> Puromycin[,c(1,3)]
```

	conc	state
1	0.02	treated
2	0.02	treated
3	0.06	treated
4	0.06	treated
5	0.11	treated
6	0.11	treated
7	0.22	treated
8	0.22	treated
9	0.56	treated
10	0.56	treated
11	1.10	treated
12	1.10	treated
13	0.02	untreated
14	0.02	untreated
15	0.06	untreated
16	0.06	untreated
17	0.11	untreated
18	0.11	untreated
19	0.22	untreated
20	0.22	untreated
21	0.56	untreated
22	0.56	untreated
23	1.10	untreated

You can also use object names in place of numbers:

```
> rownum<-c(6,8,14)
> colnum<-2
> Puromycin[rownum,colnum]
```

```
[1] 139 152  51
```

Or even functions:

```
> Puromycin[sqrt(25),]
```

	conc	rate	state
5	0.11	123	treated

Finally, you can refer to specific entries using a combination of the variable name and bracket notation. For example, to select the tenth observation for the rate variable:

```
> Puromycin$rate[10]
```

```
[1] 201
```

You can view more information about the Puromycin dataset (or any of the other dataset included with R) with the help function:

```
> help(Puromycin)
```

The Data Editor

As an alternative to viewing datasets in the command window, R has a spreadsheet style viewer called the *data editor,* which allows you to view and edit data frames. To open the Puromycin dataset in the data editor window, use the command:

```
> fix(Puromycin)
```

Alternatively, you can select Data Editor from the Edit menu and enter the name of the dataset that you want to view when prompted. The dataset opens in the data editor window, as shown in Figure 1-2. Here you can make changes to the data. When you have finished, close the editor window to apply them.

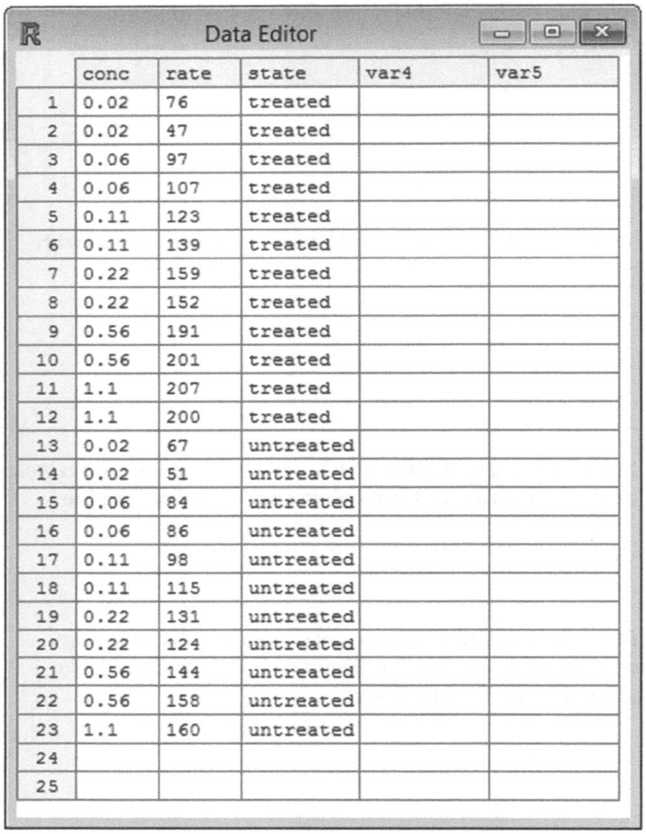

	conc	rate	state	var4	var5
1	0.02	76	treated		
2	0.02	47	treated		
3	0.06	97	treated		
4	0.06	107	treated		
5	0.11	123	treated		
6	0.11	139	treated		
7	0.22	159	treated		
8	0.22	152	treated		
9	0.56	191	treated		
10	0.56	201	treated		
11	1.1	207	treated		
12	1.1	200	treated		
13	0.02	67	untreated		
14	0.02	51	untreated		
15	0.06	84	untreated		
16	0.06	86	untreated		
17	0.11	98	untreated		
18	0.11	115	untreated		
19	0.22	131	untreated		
20	0.22	124	untreated		
21	0.56	144	untreated		
22	0.56	158	untreated		
23	1.1	160	untreated		
24					
25					

Figure 1-2. *The data editor window*

Although the data editor can be useful for making minor changes, there are usually more efficient ways of manipulating a dataset. These are covered in Chapter 3.

Workspaces

The *workspace* is the virtual area containing all of the objects you have created in the session. To see a list of all of the objects in the workspace, use the objects function:

```
> objects()
```

You can delete objects from the workspace with the rm function:

```
> rm(height, string1, string2)
```

To delete all of the objects in the workspace, use the command:

```
> rm(list=objects())
```

You can save the contents of the workspace to a file, which allows you to resume working with them at another time.

Windows users can save the workspace by selecting File then Save Workspace from the drop-down menus, then naming and saving the file in the usual way. Ensure that the file name has the .RData file name extension, as it will not be added automatically.

R automatically loads the most recently saved workspace at the beginning of each new session. You can also open a previously saved workspace by selecting File, then Open Workspace, from the drop-down menus and selecting the file in the usual way. Once you have opened a workspace, all of the objects within it are available for you to use.

Mac users can find options for saving and loading the workspace from the Workspace menu.

Linux users can save the workspace by entering the command:

```
> save.image("/home/Username/folder/filename.RData")
```

The file path can be either absolute or relative to the home directory.

To load a workspace, use the command:

```
> load("/home/Username/folder/filename.RData")
```

Error Messages

Sometimes R will encounter a problem while trying to complete one of your commands. When this happens, a message is displayed in the console window to inform you of the problem. These messages come in two varieties, known as *error messages* and *warning messages*.

Error messages begin with the text Error: and are displayed when R is not able to perform the command at all.

One of most common causes of error messages is giving a command that is not a valid R command because it contains a symbol that R does not understand, or because a symbol is missing or in the wrong place. This is known as a *syntax error*. In the following example, the error is caused by an extra closing parenthesis at the end of the command:

```
> round(3.141592))
```

```
Error: unexpected ')' in "round(3.141592))"
```

Another common cause of errors is mistyping an object name so that you are referring to an object that does not exist. Remember that object names are case-sensitive:

```
> log(object5)
```

```
Error: object 'object5' not found
```

The same applies to function names, which are also case-sensitive:

```
> Log(3.141592)
```

```
Error: could not find function "Log"
```

A third common cause of errors is giving the wrong type of input to a function, such as a data frame where a vector is expected, or a character string where a number is expected:

```
> log("Hello!")
```

```
Error in log("Hello!") : Non-numeric argument to mathematical function
```

Warning messages begin with the text Warning: and tell you about issues that have not prevented the command from being completed, but that you should be aware of. For example, the command below calculates the natural logarithm of each of the values in the temperatures vector. However, the logarithm cannot be calculated for all of the values, as some of them are negative:

```
> log(temperatures)
```

```
[1] 1.0986123 1.3244190       NaN 0.1823216       NaN
Warning message:
In log(temperatures) : NaNs produced
```

Although R is still able to perform the command and produce output, it displays a warning message to draw to your attention to this issue.

Script Files

A script file is a type of text file that allows you to save your commands so that they can be easily reviewed, edited, and repeated.

To create a new script file, select New Script from the File menu. Mac users should select New Document from the File menu. This opens a new R Editor window where you can type commands. The Linux version of R does not include a script editor; however, a number of external editors are available. To see a list of these, go to www.sciviews.org/_rgui/projects/Editors.html.

To run a command from the R Editor in the Windows environment, place the cursor on the line that you want to run, then right-click and select Run Line or Selection. You can also use the shortcut Ctrl+R. Alternatively, you can click the run button, which looks like this:

To run several commands, highlight a selection of commands then right-click and select Run Line or Selection, as shown in Figure 1-3.

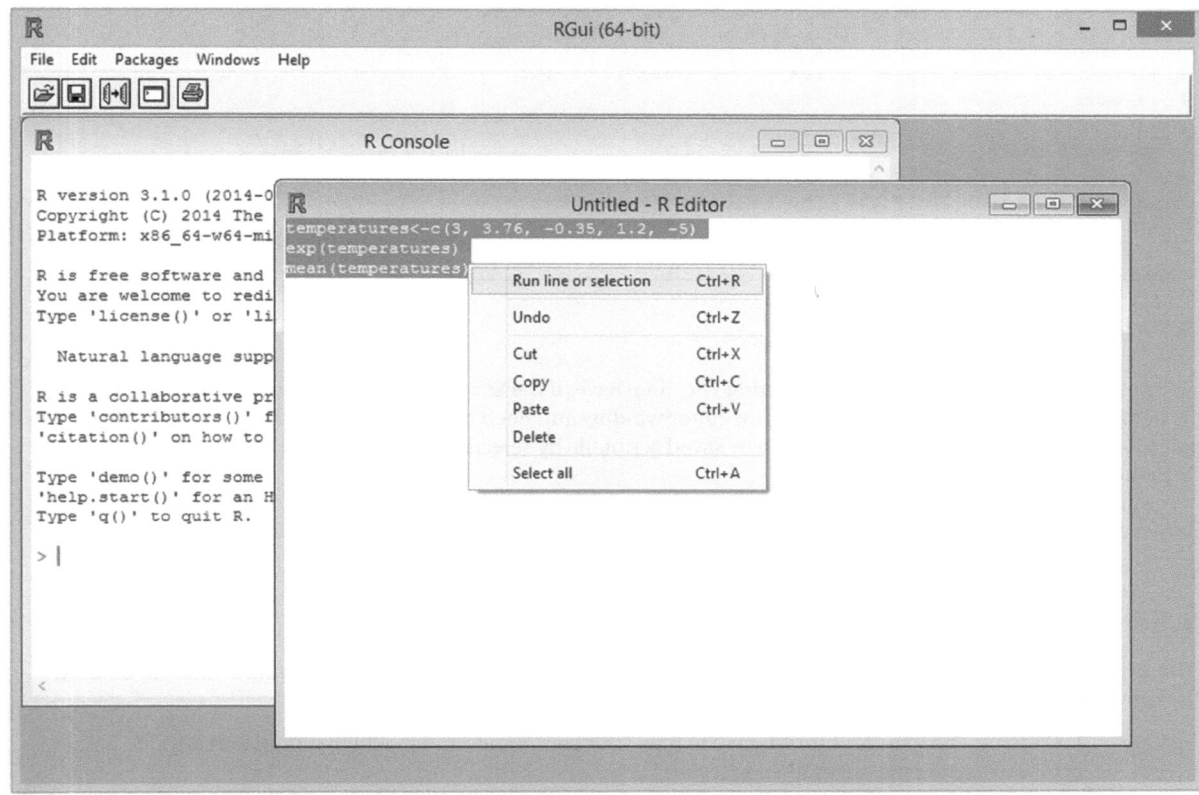

Figure 1-3. *Running commands from a script file in the Windows environment*

Mac users can run the current line or a selection of commands by pressing Cmd+Return.

Once you have run the selected commands, they are submitted to the command window and executed one after the other.

If your script file is going to be used by someone else or if you are likely to return to it after a long time, it is helpful to add some *comments*. Comments are additional text that are not part of the commands themselves but are used to make notes and explain the commands.

Add comments to your script file by typing the hash sign (#) before the comment. R ignores any text following a hash sign for the remainder of the line. This means that if you run a section of commands that has comments in, the comments will not cause errors. Figure 1-4 shows a script file with comments.

```
R                 Untitled - R Editor          □  ◻  ✕
# Script for creating a vector of temperatures and
# calculating the mean temperature

temperatures<-c(3, 3.76, -0.35, 1.2, -5)
# exp(temperatures) This line is commented out
mean(temperatures) # This line calculates the mean
```

Figure 1-4. *Script file with comments*

You can save a script file using by selecting File, then Save. If the Save option is not shown in the File menu, it is because you don't have focus on the script editor window and need to select it. The file is given the *.R* file name extension. Similarly, you can open a previously saved script file by selecting File, then Open Script, and selecting the file in the usual manner.

Mac users can save a script file by selecting the icon in the top left-hand corner of the script editor window. They can open a previously saved script file by selecting Open Document from the File menu.

Summary

The purpose of this chapter is to familiarize you with the R interface and the programming terms that will be used throughout the book. Make sure that you understand the following terms before proceeding:

- **R Console** The window into which you type your commands and in which output and any error or warning messages are displayed.

- **Command** A typed instruction to R.

- **Command prompt** The symbol used by R to indicate that it is ready to receive your command, which looks like this: >.

- **Function** A set of commands that have been given a name and together perform a specific task.

- **Argument** A value or piece of data supplied to a function as input.

- **Object** A piece of data or information that has been stored and given a name.

- **Vector** An object that contains several data values of the same type arranged in a particular order.

- **Data frame** A type of object that is suitable for holding a dataset.

- **Workspace** The virtual area containing all of the objects created in the session, which can be saved to a file with the .RData file name extension.

- **Script file** A file with the .R extension, which is used to save commands and comments.

Now that you are familiar with the R interface, we can move on to Chapter 2 where you will learn how to get your data into R.

CHAPTER 2

■ ■ ■

Working with Data Files

Before you can begin any statistical analysis, you will need to learn to work with external data files so that you can import your data. R is able to read the comma-separated values (CSV), tab-delimited, and data interchange format (DIF) file formats, which are some of the standard file formats commonly used to transfer data between statistical and database packages. With the help of an add-on package called foreign, it is possible to import a wider range of file types.

Whether you have personally recorded your data on paper or in a spreadsheet, or received a data file from someone else, this chapter will explain how to get your data into R.

You will learn how to:

- enter your data by typing the values in directly

- import plain text files, including the CSV, tab-delimited, and DIF file formats

- import Excel® files by first converting them to the CSV format

- import a dataset stored in a file type specific to another software package such as an SPSS or Stata data file

- work with relative file paths

- export a dataset to a CSV or tab-delimited file

Entering Data Directly

If you have a small dataset that is not already recorded in electronic form, you may want to input your data into R directly.

Consider the dataset shown in Table 2-1, which gives some data for four U.K. supermarkets chains. It is representative of a typical dataset in which the columns represent variables and each row holds one observation.

Table 2-1. *Data for the U.K.'s Four Largest Supermarket Chains (2011); see Appendix C for more details*

Chain	Stores	Sales Area (1,000 sq ft)	Market Share %
Morrisons	439	12261	12.3
Asda			16.9
Tesco	2715	36722	30.3
Sainsbury's	934	19108	16.5

To enter a dataset into R, the first step is to create a vector of data values for each variable using the c function, as explained under "Vectors" in Chapter 1. So, for the supermarkets data, input the four variables:

```
> Chain<-c("Morrisons", "Asda", "Tesco", "Sainsburys")
> Stores<-c(439, NA, 2715, 934)
> Sales.Area<-c(12261, NA, 36722, 19108)
> Market.Share<-c(12.3, 16.9, 30.3, 16.5)
```

The vectors should all have the same length, meaning that they should contain the same number of values. Where a data value is missing, enter the characters NA in its place. Remember to put quotation marks around non-numeric values, as shown for the Chain variable.

Once you have created vectors for each of the variables, use the data.frame function to combine them to form a data frame:

```
> supermarkets<-data.frame(Chain, Stores, Sales.Area, Market.Share)
```

You can check the dataset has been entered correctly by entering its name:

```
> supermarkets
```

	Chain	Stores	Sales.Area	Market.Share
1	Morrisons	439	12261	12.3
2	Asda	NA	NA	16.9
3	Tesco	2715	36722	30.3
4	Sainsburys	934	19108	16.5

After you have created the data frame, the individual vectors still exist in the workspace as separate objects from the data frame. To avoid any confusion, you can delete them with the rm function:

```
> rm(Chain, Stores, Sales.Area, Market.Share)
```

Importing Plain Text Files

The simplest way to transfer data to R is in a plain text file, sometimes called a *flat text file*. These are files that consist of plain text with no additional formatting and can be read by plain text editors such as Microsoft Notepad, TextEdit (for Mac users), or gedit (for Linux users). There are several standard formats for storing spreadsheet data in text files, which use symbols to indicate the layout of the data. These include:

- Comma-separated values or comma-delimited (.csv) files
- Tab-delimited (.txt) files
- Data interchange format (.dif) files

These file formats are useful because they can be read by all of the popular statistical and database software packages, allowing easy transfer of datasets between them.

The following subsections explain how you can import datasets from these standard file formats into R.

CSV and Tab-Delimited Files

Comma-separated values (CSV) files are the most popular way of storing spreadsheet data in a plain text file. In a CSV file, the data values are arranged with one observation per line and commas are used to separate data value within each line (hence the name). Sometimes semicolons are used instead of commas, such as when there are commas within the data itself. The tab-delimited file format is very similar to the CSV format except that the data values are separated with horizontal tabs instead of commas. Figures 2-1 and 2-2 show how the supermarkets data looks in the CSV and tab-delimited formats. These files are available with the downloads for the book (www.apress.com/9781484201404).

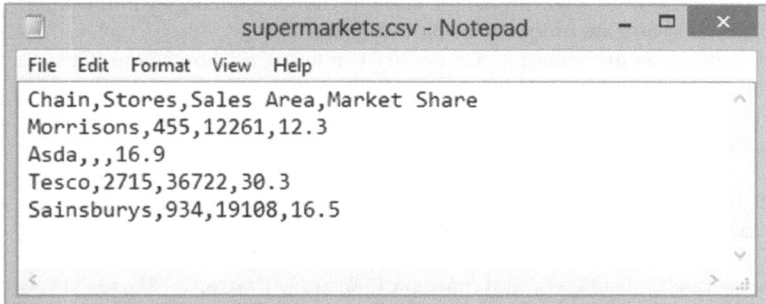

Figure 2-1. *The supermarkets dataset saved in the CSV file format*

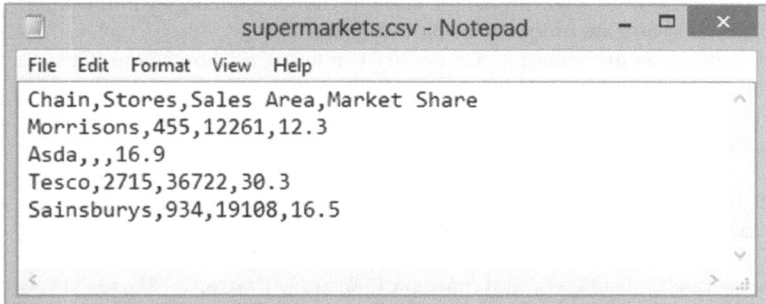

Figure 2-2. *The supermarkets dataset saved in the tab-delimited file format*

You can import a CSV file with the read.csv function:

```
> dataset1<-read.csv("C:/folder/filename.csv")
```

This command imports data from the file location C:/folder/filename.csv and saves it to a data frame called dataset1.

Remember that when choosing a dataset name, the usual object naming rules apply (see the "Objects" section in Chapter 1). It is a good idea to choose a short but meaningful name that describes the data in the file.

The file path C:/folder/filename.csv is the location of the CSV file on your hard drive, network drive, or storage device. Notice that you must use forward slashes to indicate directories instead of the back slashes used by Windows, and that you must enclose the file path between quotation marks. For Mac and Linux users, the file path will begin with a forward slash, for example, /Users/username/folder/filename.csv.

If the dataset has been successfully imported, there is no output and R displays the command prompt once it has finished importing the file. Otherwise, you will see an error message explaining why the import failed. Assuming there are no issues, your data will be stored in the data frame named dataset1. You can view and check the data by typing the dataset name at the command prompt, or by opening it with the data editor.

For importing tab-delimited files, there is a similar function called read.delim:

```
> dataset1<-read.delim("C:/folder/filename.txt")
```

When you use the read.csv or read.delim functions to import a file, R assumes that the entries in the first line of the file are the variable names for the dataset. Sometimes R will adjust the variable names so that they follow the naming rules (see the "Objects" section in Chapter 1) and are unique within the dataset.

If your file does not contain any variable names, set the header argument to F (for false), as shown here. This prevents R from using the first line of your data as the variable names:

```
> dataset1<-read.csv("C:/folder/filename.csv", header=F)
```

When you set the header argument to F, R assigns generic variable names of V1, V2, and so on. Alternatively, you can supply your own names with the col.names argument:

```
> dataset1<-read.csv("C:/folder/filename.csv", header=F, col.names=c("Name1", "Name2", "Name3"))
```

When using the col.names argument, make sure that you give the same number of names as there are variables in the file. Otherwise, you will either see an error message or find that some of the variables are left unnamed.

In a CSV or tab-delimited file, missing data is usually represented by an empty field. However, some files may use a symbol or character such as a decimal point, the number 9999, or the word NULL as a place holder. If so, use the na.strings argument to tell R which characters to interpret as missing data:

```
> dataset1<-read.csv("C:/folder/filename.csv", na.strings=".")
```

Note that R automatically interprets the letters NA as indicating missing data. Blank spaces found in numeric variables (but not character variables) are also recognized as missing data.

In some parts of the world, a comma is used instead of a dot to denote the decimal point in numbers. Figure 2-3 shows a tab-delimited file with numbers in this format. R has two special functions for dealing with data in this form, called read.csv2 and read.delim2, which you can use in place of the read.csv and read.delim functions.

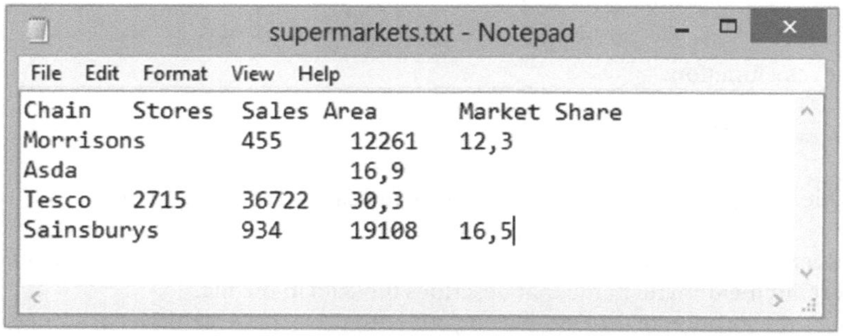

Figure 2-3. The supermarkets dataset saved in the tab-delimited file format, with commas used to represent the decimal point

DIF Files

To import a data interchange format (DIF) file with the `.dif` file name extension, use the `read.DIF` function:

```
> dataset1<-read.DIF("C:/folder/filename.dif")
```

By default, the `read.DIF` function assumes that there are no variable names in the file (unlike the `read.csv` and `read.delim` functions). If the file does contain variable names, set the `header` argument to T (for true), as shown below. Otherwise, R will treat the variable names as part of the data values:

```
> dataset1<-read.DIF("C:/folder/filename.dif", header=T)
```

As when importing CSV and tab-delimited files, you can use the `na.strings` argument to tell R about any values that it should interpret as missing data:

```
> dataset1<-read.DIF("C:/folder/filename.dif", na.string="NULL")
```

Note that if you are importing a DIF file that was created by Microsoft Excel, you may see the error message `'More rows than specified in header'; maybe use 'transpose=TRUE'` when you try to import the file. This error message appears because some versions of Microsoft Excel use a modified version of the DIF file format instead of the standard one. You can resolve the problem by setting the `transpose` argument to T:

```
> dataset1<-read.DIF("C:/folder/filename.dif", transpose=T)
```

Other Plain Text Files

As well as all of the functions available for importing specific file formats, R also has a generic function for importing data from plain text files called `read.table`. It allows you to import any plain text file in which the data is arranged with one observation per line.

Consider the file shown in Figure 2-4, which has data arranged with one observation per line and data values separated by the forward slash symbol (/).

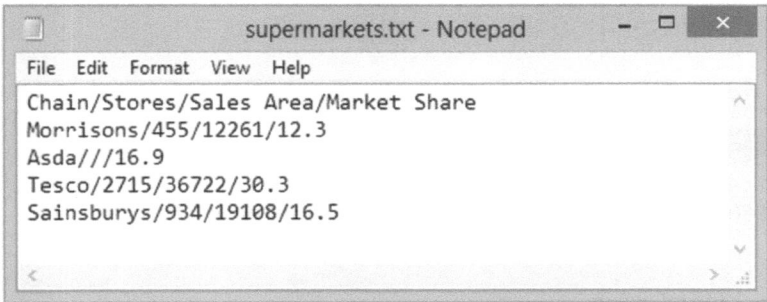

Figure 2-4. *The supermarkets data in a nonstandard file type*

You could import the file with this command:

```
> dataset1<-read.table("C:/folder/supermarkets.txt", sep="/", header=T)
```

By default, the read.table function assumes that there are no variable names in the first row of the file (unlike the read.csv and read.delim functions). If the file has variable names in the first row (as in this example), set the header argument to T.

As with the other import functions, you can use the na.strings argument to tell R of any values to interpret as missing data:

```
dataset1<-read.table("C:/folder/filename.txt", sep="/", header=T, na.strings="NULL")
```

Importing Excel Files

The simplest way to import a Microsoft Excel file is to save your Excel file as a CSV file, which you can then import, as explained earlier in this chapter under "CSV and Tab-Delimited Files."

First open your file in Excel and ensure that the data is arranged correctly within the spreadsheet, with one variable per column and one observation per row. If the dataset includes variable names, then these should be placed in the first row of the spreadsheet. Otherwise, the data values should begin on the first row. Figure 2-5 shows how the supermarkets data looks when correctly arranged in an Excel file.

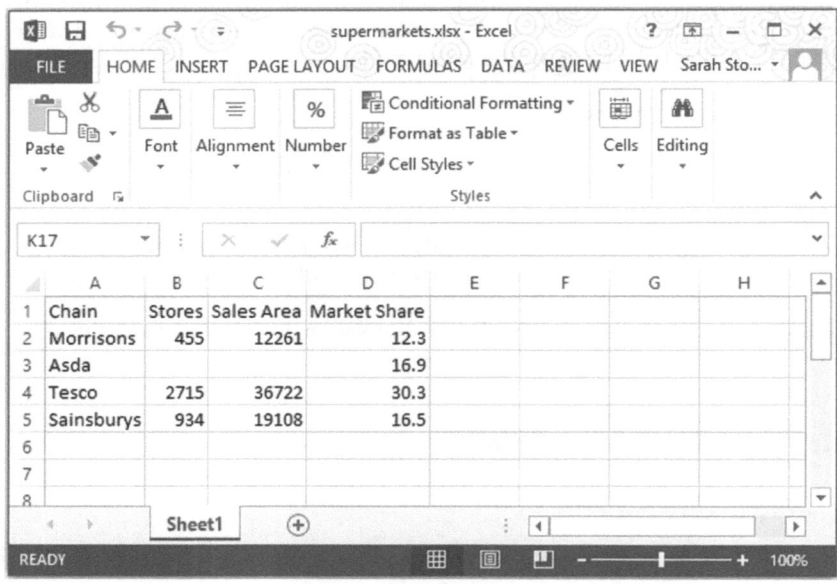

Figure 2-5. *The correct way to arrange a dataset in an Excel spreadsheet, to facilitate easy conversion to the CSV file format*

To ensure a smooth file conversion, check the following:

- There are no empty cells above or to the left of the data grid
- There are no merged cells
- There is no more than one row of column headers
- There is no formatting such as bold, italic or colored text, cell borders or background colors
- Where data values are missing, the cell is left empty

- There are no commas in large numbers (e.g., 1324157 is acceptable but 1,324,157 is not)

- If exponential (scientific) notation is used, the format is correct (e.g., 0.00312 can be expressed as 3.12e-3 or 3.12E-3)

- There are no currency, unit, or percent symbols in numeric variables (symbols in categorical variables or in the variable names are fine)

- The minus sign is used to indicate negative numbers (e.g., -5) and not brackets (parentheses) or red text

- The workbook has only one worksheet

When the data is prepared, save the spreadsheet as a CSV file by selecting Save As from the File menu. When the Save As dialog box appears, select the .csv file type from the Save As Type field, as shown in Figure 2-6. Then save the file as usual. Excel will give you a warning that all formatting will be lost, which you can accept.

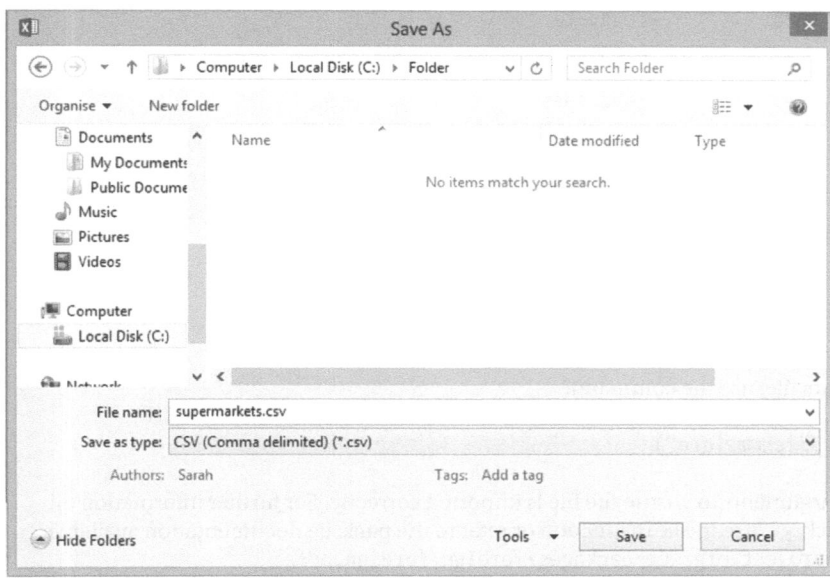

Figure 2-6. *Saving an Excel file as a CSV file*

The CSV file is now ready for you to import with the read.csv function, as explained previously in this chapter (see "CSV and Tab-Delimited Files").

If you do not have access to Excel, you can use an add-on package such as xlsx or xlsReadWrite to import Excel files directly. See Appendix B for more details on using add-on packages.

Importing Files from Other Software

Sometimes you may need to import a dataset that is saved in a file format specific to another statistical package, such as an SPSS or Stata data file.

If you have access to the software, the simplest solution is to open the file using the software and convert the file to the CSV file format using the Save As or Export option, which is usually found in the File menu. Once the file is in CSV format, you can import it with the read.csv function (see "CSV and Tab-Delimited Files" earlier in this chapter).

If you are not able to convert the file, then you can use an add-on package called `foreign`, which allows you to directly import data from files types produced by some of the popular statistical software packages.

Add-on packages are covered in greater detail in Appendix A. For now, you just need to know that an add-on package contains additional functions that are not part of the standard R installation. To use the functions within the `foreign` package, you must first load the package.

To load the `foreign` package, select Load Package from the Packages menu. When the list of packages appears, select "foreign" and press OK. Once the package has loaded, all of the functions within it will be available for you to use for the duration of the session.

Table 2-2 lists some of the functions available for importing foreign file types.

Table 2-2. *Some of the Functions Available in the Foreign Add-on Package*

File type	Extension	Function
Database format file	.dbf	read.dbf
Stata versions 5 to 12 data file	.dta	read.dta
Minitab portable worksheet file	.mtp	read.mtp
SPSS data file	.sav	read.spss
SAS transfer format	.xport	read.xport
Epi Info data file	.rec	read.epiinfo
Octave text data file	.txt	read.octave
Attribute-relation file	.arff	read.arff
Systat file	.sys, .syd	read.systat

For example, to import a Stata data file, use the command:

```
> dataset1<-read.dta("C:/folder/filename.dta")
```

You may need to use additional arguments to ensure the file is imported correctly. For further information on using the functions in the `foreign` package, use the `help` function or refer to the package documentation available from the R project website at `cran.r-project.org/web/packages/foreign/foreign.pdf`.

Using Relative File Paths

So far, we have only used *absolute* file paths to describe the location of a data file. An absolute file path gives the full address of the file, which in the Windows environment begins with a drive name such as `C:/`.

You can also use *relative* file paths, which describe the location of the file in relation to the *working directory*. The working directory is the directory in which R is set to look when given relative file paths. This is useful if you need to import or export a large number of files and don't want to type the full file path each time. To see which is the current working directory, use the command:

```
> getwd()
```

If you are using a fresh installation of R for Windows, the working directory will be your Documents folder, and R will output something like this:

```
[1] "C:/Users/Username/Documents"
```

For Mac and Linux users, the default working directory will be your home directory and will look something like this:

```
[1] "/Users/Username"
```

So, to import a CSV file that has the absolute file path:

```
C:/Users/Username/Documents/Data/filename.csv
```

you could use the command:

```
> read.csv("Data/filename.csv")
```

To change the working directory to wherever you prefer to store your data files, use the setwd function, as shown below. The change is applied for the remainder of the session:

```
> setwd("C:/folder/subfolder")
```

Exporting Datasets

R allows you to export datasets from the R workspace to the CSV and tab-delimited file formats.

To export a data frame named dataset to a CSV file, use the write.csv function:

```
> write.csv(dataset, "filename.csv")
```

For example, to export the Puromycin dataset to a file named puromycin_data.csv, use the command:

```
> write.csv(Puromycin, "puromycin_data.csv")
```

This command creates the file and saves it to your working directory (see the preceding section, "Using Relative File Paths," for how to find and set the working directory). To save the file somewhere other than in the working directory, enter the full path for the file:

```
> write.csv(dataset, "C:/folder/filename.csv")
```

If a file with your chosen name already exists in the specified location, R overwrites the original file without giving a warning. You should check the files in the destination folder beforehand to make sure that you are not overwriting anything important.

The write.table function allows you to export data to a wider range of file formats, including tab-delimited files. Use the sep argument to specify which character should be used to separate the values. To export a dataset to a tab-delimited file, set the sep argument to "\t" (which denotes the tab symbol):

```
> write.table(dataset, "filename.txt", sep="\t")
```

By default, the write.csv and write.table functions create an extra column in the file containing the observation numbers. To prevent this, set the row.names argument to F:

```
> write.csv(dataset, "filename.csv", row.names=F)
```

With the write.table function, you can also prevent variable names being placed in the first line of the file with the col.names argument:

```
> write.table(dataset, "filename.txt", sep="\t", col.names=F)
```

Summary

You should now be able to get your data into R, whether by entering it directly or by importing it from an external data file. You should also understand how to use relative file paths and be able to export a dataset to an external file.

This table summarizes the most import commands covered in this chapter.

Task	Command
Create data frame	dataset<-data.frame(vector1, vector2, vector3)
Import CSV file	dataset<-read.csv("filepath")
Import tab-delimited file &	dataset<-read.delim("filepath")
Import DIF file	dataset<-read.DIF("filepath")
Import other text file	dataset<-read.table("filepath, sep="?")
Display working directory	getwd()
Change working directory	setwd("C:/folder/subfolder")
Export dataset to CSV file	write.csv(dataset, "filename.csv")
Export dataset to tab-delimited file	write.table(dataset, "filename.txt",sep="\t")

Now that you have learned how to get your dataset into R, we can move on to Chapter 3, which explains how to prepare your dataset for analysis.

CHAPTER 3

■ ■ ■

Preparing and Manipulating Your Data

After you have imported your dataset, it is likely that you will need to make some changes before beginning any statistical analysis. You may require some new variables for your analysis, or there may be some irrelevant data that needs to be removed. Additionally, you may want to ensure that variables and categories are correctly named so that they look more presentable on any statistical output that you create. This chapter explains how you can make these types of changes to a dataset.

You will learn how to:

- rename, rearrange, and remove variables

- change the data type of variables

- calculate new variables from old ones

- divide numeric variables into categories

- modify category names for categorical (factor) variables

- manipulate character strings

- work with dates and times

- add or remove observations

- select a subset of data, either by type or as a random sample

- sort the data

More complex changes, such as combining two datasets or changing the structure of the data, are covered in Chapter 4.

This chapter uses the people dataset shown in Figure 3-1 for demonstration purposes. This dataset gives the eye color (brown, blue, or green), height in centimeters, hand span in millimeters, sex (1 for male, 2 for female), and handedness (L for left-handed, R for right-handed) of sixteen people.

	Subject	Eye.Color	Height	Hand.Span	Sex	Handedness
1	1	Brown	186	210	1	R
2	2	Green	182	220	1	R
3	3	Brown	147	167	2	
4	4	Green	157	180	2	L
5	5	Brown	170	193	1	R
6	6	Blue	169	190	2	L
7	7	brown	174	217	1	R
8	8	Blue	173	211	1	R
9	9	Blue	166	193	2	R
10	10	Blue	166	178	2	R
11	11	Brown	163	223	1	R
12	12	Blue	184	225	1	R
13	13	Blue	176	214	1	
14	14	Blue	183	218	1	R
15	15	Green	160	190	2	
16	16	Brown	173	196	1	R

Figure 3-1. *The people dataset*

This chapter also uses the pulserates, fruit, flights, customers, and coffeeshop datasets, which are all available with the downloads for this book (www.apress.com/9781484201404) in CSV format or in an R workspace file. For more information about these datasets, see Appendix C.

Variables

If your dataset has a large number of variables, you can make it more manageable by removing any unnecessary variables and arranging the remaining variables in a meaningful order. You should check that each variable has an appropriate name and an appropriate class for the type of data that it holds, as explained in the following sections.

Rearranging and Removing Variables

You can rearrange or remove the variables in a dataset with the subset function. Use the select argument to choose which variables to *keep* and in which order. Remove unwanted variables by excluding them from the list.

For example, this command removes the Subject, Height and Handedness variables from the people dataset, and rearranges the remaining variables so that Hand.Span is first, followed by Sex then Eye.Color:

```
> people1<-subset(people, select=c(Hand.Span, Sex, Eye.Color))
```

Figure 3-2 shows how the new dataset looks after the changes have been applied.

	Hand.Span	Sex	Eye.Color
1	210	1	Brown
2	220	1	Green
3	167	2	Brown
4	180	2	Green
5	193	1	Brown
6	190	2	Blue
7	217	1	brown
8	211	1	Blue
9	193	2	Blue
10	178	2	Blue
11	223	1	Brown
12	225	1	Blue
13	214	1	Blue
14	218	1	Blue
15	190	2	Green
16	196	1	Brown

Figure 3-2. *The* people1 *dataset, created by removing variables from the* people *dataset with the* subset *function*

Notice that the command creates a new dataset called people1, which is a modified version of the original, and leaves the original dataset unchanged. Alternatively, you can overwrite the original dataset with this modified version:

```
> people<-subset(people, select=c(Hand.Span, Sex, Eye.Color))
```

The subset function does more than remove and rearrange variables. You can also use it to select a subset of observations from a dataset, which is explained later in this chapter under "Selecting a subset according to selection criteria".

Another way of removing variables from a dataset is with bracket notation. This is particularly useful if you have a dataset with a large number of variables and you only want to remove a few. For example, to remove the first, third, and sixth variables from the people dataset, use the command:

```
> people1<-people[-c(1,3,6)]
```

Similarly, to retain the second, fourth, and first variables and reorder them, use the command:

```
> people1<-people[c(2,4,1)]
```

■ **Note**   See Chapter 1 under "Data Frames" for more details on using bracket notation.

Renaming Variables

The names function displays a list of the variable names for a dataset:

```
> names(people)
```

```
[1] "Subject"    "Eye.Color" "Height"     "Hand.Span"  "Sex"        "Handedness"
```

You can also use the names function to rename variables. This command renames the fifth variable in the people dataset:

```
> names(people)[5]<-"Gender"
```

Similarly, to rename the second, fourth, and fifth variables:

```
> names(people)[c(2,4,5)]<-c("Eyes", "Span.mm", "Gender")
```

Alternatively you can rename all of the variables in the dataset simultaneously:.

```
> names(people)<-c("Subject", "Eyes", "Height.cm", "Span.mm", "Gender", "Hand")
```

Make sure that you provide the same number of variable names as there are variables in the dataset.

Variable Classes

Each of the variables in a dataset has a *class*, which describes the type of data the variable contains. You can view the class of a variable with the class function:

```
> class(dataset$variable)
```

To check the class of all the variables simultaneously, use the command:

```
> sapply(dataset, class)
```

A variable's class determines how R will treat the variable when you use it in statistical analysis and plots. There are many possible variable classes in R, but only a few that you are likely to use:

> **numeric** variables contain *real* numbers, meaning positive or negative numbers with or without a decimal point. They can also contain the missing data symbol (NA)

> **integer** variables contain positive or negative numbers without a decimal point. This class behaves in much the same way as the numeric class. An integer variable is automatically converted to a numeric variable if a value with a fractional part is included

> **factor** variables are suitable for categorical data. Factor variables generally have a small number of unique values, known as *levels*. The actual values can be either numbers or character strings

> **date & POSIXlt** variables contain dates or date-times in a special format, which is convenient to work with

character variables contain character strings. A character string is any combination of unicode characters including letters, numbers, and symbols. This class is suitable for any data that does not belong to one of the other classes, such as reference numbers, labels, and text, giving additional comments or information

When you import a data file using a function such as read.csv, R automatically assigns each variable a class based on its contents. If a variable contains only numbers, R assigns the numeric or integer class. If a variable contains any non-numeric values, it assigns the factor class.

Because R does not know how you intend to use the data contained in each variable, the classes that it assigns to them may not always be appropriate. To illustrate, consider the Sex variable in the people dataset. Because the variable contains whole numbers, R automatically assigns the integer class when the data is imported. But the factor class would be more appropriate, as the values represent categories rather than counts or measurements.

You can change the class of a variable to factor with the as.factor function:

```
> dataset$variable<-as.factor(dataset$variable)
```

If you have a variable containing numeric values that for some reason has been assigned another class, you can change it using the as.numeric function. Any non-numeric values are treated as missing data and replaced with the missing data code (NA):

```
> dataset$variable<-as.numeric(dataset$variable)
```

If R has not automatically recognized a variable as numeric when importing a dataset, then it is because the variable contains at least one non-numeric value. It is wise to determine the cause, as it may be that a value has been entered incorrectly or that a symbol used to represent missing data has not been recognized.

You can change the class of a variable to character using the as.character function:

```
> dataset$variable<-as.character(dataset$variable)
```

There is also an as.Date function for creating date variables, which you will learn more about in "Working with dates and times" later in this chapter.

Calculating New Numeric Variables

You can create a new variable within a dataset in the same way that you would create any other new object, using the assignment operator (<-). So to create a new variable named var2 that is a copy of an existing variable named var1, use the command:

```
> dataset$var2<-dataset$var1
```

You can create new numeric variables from combinations of existing numeric variables and arithmetic operators and functions. For example, the command below adds a new variable called Height.Inches to the people dataset, which gives the subject's heights in inches rounded to the nearest inch:

```
> people$Height.Inches<-round(people$Height/2.54)
```

You can use bracket notation to make conditional changes to a variable. For example, to set all values of Height less than 150 cm to missing, use the command:

```
> people$Height[people$Height<150]<-NA
```

■ **Note**   You will learn more about using conditions in "Selecting a Subset According to Selection Criteria" later in this chapter and in Appendix B.

You may want to create a new variable that is a statistical summary of several of the existing variables. The apply function allows you to do this.

Consider the pulserates dataset shown in Figure 3-3, which gives pulse rate data for four patients. The patients' pulse rates are measured in triplicate and stored in variables Pulse1, Pulse2, and Pulse3.

	Patient	Pulse1	Pulse2	Pulse3
1	3051	70	77	73
2	3052	65	66	56
3	3053	64	58	60
4	3054	53	58	56

Figure 3-3. pulserates *dataset giving the pulse rates of four patients, measured in triplicate (see Appendix C for more details)*

Suppose that you want to calculate a new variable giving the mean pulse for each patient. You can create the new variable (shown in Figure 3-4) with the command:

```
> pulserates$Mean.Pulse<-apply(pulserates[2:4], 1, mean)
```

	Patient	Pulse1	Pulse2	Pulse3	Mean.Pulse
1	3051	70	77	73	73.33333
2	3052	65	66	56	62.33333
3	3053	64	58	60	60.66667
4	3054	53	58	56	55.66667

Figure 3-4. pulserates *dataset with the new* Mean.Pulse *variable*

Notice that bracket notation is used to select column numbers 2 to 4. (The "Data frames" section in Chapter 1 gives more details on using bracket notation.)

The second argument allows you to specify whether the summary should be calculated for each row (by setting it to 1) or each column (by setting it to 2). To create a new variable, set it to 1.

You can substitute the mean function with any univariate statistical summary function that gives a single value as output, such as sd or max. Table 5-1 gives a list of these (use only those marked with an asterisk).

Dividing a Continuous Variable into Categories

Sometimes you may want to create a new categorical variable by classifying the observations according to the value of a continuous variable.

For example, consider the people dataset shown in Figure 3-1 Suppose that you want to create a new variable called Height.Cat, which classifies the people as "Short", "Medium", and "Tall" according to their height. People less than 160 cm tall are classified as Short, people between 160 cm and 180 cm tall are classified as Medium, and people greater than 180 cm tall are classified as Tall.

You can create the new variable with the cut function:

```
> people$Height.Cat<-cut(people$Height, c(150, 160, 180, 200), c("Short", "Medium", "Tall"))
```

Figure 3-5 shows the people dataset with the new Height.Cat variable.

	Subject	Eye.Color	Height	Hand.Span	Sex	Handedness	Height.Cat
1	1	Brown	186	210	1	R	Tall
2	2	Green	182	220	1	R	Tall
3	3	Brown	147	167	2		
4	4	Green	157	180	2	L	Short
5	5	Brown	170	193	1	R	Medium
6	6	Blue	169	190	2	L	Medium
7	7	brown	174	217	1	R	Medium
8	8	Blue	173	211	1	R	Medium
9	9	Blue	166	193	2	R	Medium
10	10	Blue	166	178	2	R	Medium
11	11	Brown	163	223	1	R	Medium
12	12	Blue	184	225	1	R	Tall
13	13	Blue	176	214	1		Medium
14	14	Blue	183	218	1	R	Tall
15	15	Green	160	190	2		Short
16	16	Brown	173	196	1	R	Medium

Figure 3-5. *The people dataset with the new Height.Cat variable*

When using the cut function, the numbers of group boundaries (in this example four) must be one more than the number of group names (in this example three). If a data value is equal to one of the boundaries, it is placed in the category below. Make sure your categories cover the whole range of the data values; otherwise, the new variable will have missing values. In this example, there is one observation (subject 3) that does not fall in to any of the categories that have been defined, so has a missing value for the Height.Cat variable.

If you prefer, you can specify the number of categories and let R determine where the boundaries should be. R divides the range of the variable to create evenly sized categories. For example, this command shows how you would split the Height variable into three evenly sized categories:

```
> people$Height.Cat<-cut(people$Height, 3, c("Short", "Medium", "Tall"))
```

Any variables you create with the cut function are automatically assigned the factor class.

■ **Note** Always consider carefully whether you really need to divide a numeric variable into categories. Numeric variables contain more information than categorical variables, so it is often wisest to include the original numeric variable directly in your statistical models where possible.

Working with Factor Variables

As explained under "Variable classes," factor variables are suitable for holding categorical data. To change the class of a variable to factor, use the as.factor function:

```
> people$Sex<-as.factor(people$Sex)
```

A factor variable has a number of levels, which are all of the unique values that the variable takes (i.e., all of the possible categories). To view the levels of a factor variable, use the levels function:

```
> levels(people$Sex)
```

```
[1] "1" "2"
```

Because the level names will appear on any plots and statistical output that you create based on the variable, it is helpful if they are meaningful and attractive. You can change the names of the levels:

```
> levels(people$Sex)<-c("Male", "Female")
```

You must give the same number of names as there are levels of the factor, and enter the new names in corresponding order.

You can also combine factor levels by renaming them. Consider the Eye.Color variable in the people dataset. Using the levels function, you can see that there is an extra level resulting from a spelling variation:

```
> levels(people$Eye.Color)
```

```
[1] "Blue" "brown" "Brown" "Green"
```

To rename the second factor level so that it has the correct spelling, use the command:

```
> levels(people$Eye.Color)[2]<-"Brown"
```

When the factor levels are viewed again, you can see that the two levels have been combined:

```
> levels(people$Eye.Color)
```

```
[1] "Blue"  "Brown" "Green"
```

You can change the order of the levels with the relevel function. For example, to make Brown the first level of the Eye.Color variable, use the command:

```
> people$Eye.Color<-relevel(people$Eye.Color, "Brown")
```

The order of the factor levels is important, because if you include the factor in a statistical model, R uses the first level of the factor as the *reference level*. You will learn more about this in Chapter 11.

Manipulating Character Variables

R has a number of functions for manipulating character strings. Three of the most useful are paste (for concatenating strings), substring (for extracting a substring), and grep (for searching a string). These are demonstrated in the following subsections.

Concatenating Character Strings

The paste function allows you to create new character variables by pasting together existing variables (of any class) and other characters.

Consider the fruit dataset shown in Figure 3-6, which gives prices for a selection of fruit.

	Product	Price	Unit
1	Apricots	10	per kg
2	Baby Watermelon	2	each
3	Bananas	0.68	per kg
4	Blush Pears	2.1	per kg
5	Braeburn Apples	1.65	per kg
6	Bramley Cooking Apples	1.55	per kg
7	Cantaloupe Melon	2	each

Figure 3-6. fruit *dataset giving U.K. fruit prices for August 2012 (see Appendix C for more details)*

Suppose that you want to create a new variable giving a price label for each of the fruit. The label should have the product description, price with pound sign, and sale unit. You can create the new variable (shown in Figure 3-7) with the command:

```
> fruit$Label<-paste(fruit$Product, ": £", format(fruit$Price, trim=T, digits=2), " ", fruit$Unit, sep="")
```

	Product	Price	Unit	Label
1	Apricots	10	per kg	Apricots: £10.00 per kg
2	Baby Watermelon	2	each	Baby Watermelon: £2.00 each
3	Bananas	0.68	per kg	Bananas: £0.68 per kg
4	Blush Pears	2.1	per kg	Blush Pears: £2.10 per kg
5	Braeburn Apples	1.65	per kg	Braeburn Apples: £1.65 per kg
6	Bramley Cooking Apples	1.55	per kg	Bramley Cooking Apples: £1.55 per kg
7	Cantaloupe Melon	2	each	Cantaloupe Melon: £2.00 each

Figure 3-7. fruits *dataset with new* Label *variable*

By default, the paste function inserts a space between each of the components being pasted together. In this example, sep="" has been added to prevent this, so that spaces are not inserted in unwanted places such as between the pound sign and the price. Instead, spaces have been inserted manually where required, placed between quotation marks. You can also use the sep argument to specify another keyboard symbol to use as a separator.

Notice that the format function has been used to ensure that the fruit prices are always displayed to two decimal places.

Extracting a Substring

The substring function allows you to create a new variable by extracting a section of characters from an existing variable.

Consider the flights dataset shown in Figure 3-8a. The Flight.Number variable gives flight numbers that begin with a two-letter prefix, indicating which airline the flight is operated by. Suppose that you wish to create two new variables, one named Airline giving the two-letter airline prefix and another named Ref giving the number component, as shown in Figure 3-8b.

	Date	Time	Flight.Number	Destination
1	12/01/2012	20:30	BE898	GLASGOW
2	12/01/2012	20:35	BE775	EDINBURGH
3	13/01/2012	06:50	BE382	DUBLIN
4	13/01/2012	07:00	BE861	MANCHESTER
5	13/01/2012	07:05	BE1011	AMSTERDAM
6	13/01/2012	08:00	AF6541	RENNES
7	13/01/2012	08:00	BE3025	RENNES

	Airline	Ref
	BE	898
	BE	775
	BE	382
	BE	861
	BE	1011
	AF	6541
	BE	3025

(a) Original dataset   *(b) New variables*

Figure 3-8. *flights dataset giving details of flights from Southampton Airport (see Appendix C for more details)*

When using the substring function, give the positions within the character string of the first and last characters you want to extract. So to extract the first two characters from the Flight.Number variable to create the Airline variable, use the command:

```
> flights$Airline<-substring(flights$Flight.Number, 1, 2)
```

You can also give a starting position only, and the remainder of the string is extracted. So to create the Ref variable, use the command:

```
> flights$Ref<-substring(flights$Flight.Number, 3)
```

Note that although the new Ref variable contains numbers, it still has the character class because it was created with the substring function. If you wanted to use it as a numeric variable, you can convert it with the as.numeric function as described under "Variable classes."

Searching a Character Variable

The grep function allows you to search a character string for a search term.

Consider the customers dataset shown in Figure 3-9. Suppose that you want to identify all of the customers who live in the city of Reading.

	Name	Address
1	Mr A. Jackson	17 Bridge Street, Reading, RG3 2QN
2	Ms D. Phillips	112 park avenue, reading, berkshire, RG21NY
3	Mrs S. O'Neill	Flat 1, 72 Norfolk Road, Maidenhead, SL6 7AZ
4	Mr D. Singh	373 Castle Lane, READING, RG5 2LL
5	Mr A. Rojas	2 Green Lane, Wokingham, RG40 2NA

Figure 3-9. *The customers dataset (see Appendix C for more details)*

This command searches the `Address` variable for the term "reading":

```
> grep("reading", customers$Address)
```

```
[1] 2
```

R outputs the number 2 to indicate that observation number 2 contains the term "reading".

Notice that R has only returned one result because the search in case sensitive, so that "Reading" and "READING" are not considered matches to the search term "reading". To change this, set the `ignore.case` argument to T:

```
> grep("reading", customers$Address, ignore.case=T)
```

```
[1] 1 2 4
```

Instead of displaying the observation numbers, you can save them to an object. This allows you to create a new dataset containing only the observations that matched the search term:

```
> matches<-grep("reading", customers$Address, ignore.case=T)
> reading.customers<-customers[matches,]
```

Working with Dates and Times

R has special date and date-time variable classes that make this type of data easier to work with. When you import a dataset, R does not automatically recognize date variables. Instead, they are assigned one of the other classes according to their contents. You can convert these variables with the `as.Date` and `strptime` functions.

For variables containing just dates (without times), use the `as.Date` function to convert the variable to the date class. The command takes the form:

```
> dataset$variable<-as.Date(dataset$variable, "format")
```

where `"format"` tells R how to read the dates. R uses a special notation for specifying the format of a date, shown in Table 3-1.

Table 3-1. *The Most Commonly Used Symbols for Date-Time Formats.*
Enter `help(strptime)` *to View a Complete List*

Symbol	Meaning	Possible values
%d	Day of the month	01 to 31, or 1 to 31
%m	Month number	01 to 12
%b	Three letter abbreviated month name	Jan, Feb, Mar, Apr, etc.
%B	Full month name	January, February, March, April, etc.
%y	Two-digit year	00 to 99, e.g., 10 for 2010
%Y	Four-digit year	e.g., 2004
%H	Hour in 24-hour format	0 to 23, e.g. 19 for 7pm

(continued)

Table 3-1. (*continued*)

Symbol	Meaning	Possible values
%M	Minute past the hour	00 to 59
%S	Seconds past the hour	00 to 59
%I	Hour in 12-hour format	01 to 12
%p	AM or PM	AM or PM

Consider the Date variable in the coffeeshop dataset, shown in Figure 3-10. The variable has dates in the format *dd/mmm/yyyy*.

	Date	Sales
1	21/NOV/2011	2144.88
2	22/NOV/2011	1702.99
3	23/NOV/2011	2731.45
4	24/NOV/2011	1943.04
5	25/NOV/2011	1862.09

Figure 3-10. *coffeeshop dataset (see Appendix C for more details)*

To convert the variable to the date class, use the command:

```
> coffeeshop$Date<-as.Date(coffeeshop$Date, "%d/%b/%Y")
```

The format %d/%b/%Y tells R to expect a day (%d), three-letter month name (%b), and four-digit year (%Y), separated by forward slashes. The format must be enclosed in quotation marks.

There are two date formats that R recognizes without you needing to specify them, which are *yyyy-mm-dd* and *yyyy/mm/dd*. If your dates are in either of these formats, then you don't need to give a format when using the as.Date function.

For variables containing dates with time information, use the strptime function to convert the variable to the POSIXlt class.

For example, suppose that you want to create a new date-time variable from the Date and Time variables in the flights dataset (refer to Figure 3-6).

Before you can create a date-time variable, you will need to combine the two variables to create a single character variable using the paste function (refer to the "Concatenating character strings" section):

```
> flights$DateTime<-paste(flights$Date, flights$Time)
```

Once the date and time are together in the same character string, you can use the strptime function to convert the variable class. The strptime function is used in the same way as the as.Date function:

```
> flights$DateTime<-strptime(flights$DateTime, "%d/%m/%Y %H:%M")
```

Once your variable has the date or POSIXlt class, you can perform a number of date-related operations using functions designed for these variable classes.

For example, to find the length of the time interval between two dates or date-times, use the `difftime` function:

```
> dataset$duration<-difftime(dataset$enddate, dataset$startdate, units="hours")
```

Options for the `units` argument are `secs`, `mins`, `hours`, `days` (the default), and `weeks`.
To compare a date variable with the current date (e.g., to calculate an age), use the `Sys.Date` function:

```
> dataset$age<-difftime(Sys.Date(), dataset$dob)
```

You can use arithmetic operators to add or subtract days (for date variables) or seconds (for `POSIXlt` variables). For example, to find the date one week before a given date:

```
> dataset$newdatevar<-dataset$datevar-7
```

To find which day of the week a date falls on, use the `weekdays` function. There are also similar functions called `months` and `quarters`:

```
> coffeeshop$Day<-weekdays(coffeeshop$Date)
```

The `round` function can also be used with date-time variables. Specify one of the time units, `secs`, `mins`, `hours`, or `days`:

```
> round(flights$DateTime, units="hours")
```

You can create a character variable from a date variable with the `format` function. Specify how you want R to display the date using the format symbols given in Table 3-1:

```
> dataset$charvar<-format(dataset$datevar, format="%d.%m.%Y")
```

Adding and Removing Observations

You can use the data editor to add new observations to a data frame, and bracket notation to remove specific observations. R has a special function called `unique` for removing duplicates.

If you want to remove all of the observations that match specified criteria or belong to a particular group, use the `subset` function as explained under "Selecting a subset according to selection criteria" later in this chapter.

Adding New Observations

The simplest way to add additional observations to a dataset is with the data editor, which you can open with the command:

```
> fix(dataset)
```

When the editor window opens, type the values for the new observation into the first empty row beneath the existing data. If any of the values are missing, leave the relevant cell empty. When you have finished adding new values, close the data editor to apply the changes.

Removing Specific Observations

The simplest way to remove individual observations from a dataset is using bracket notation (see the "Data Frames" section in Chapter 1). For example, to remove observation numbers 2, 4, and 7, use the command:

```
> dataset<-dataset[-c(2,4,7),]
```

Be careful to include the comma before the closing bracket; otherwise, you will remove columns rather than rows.

Remember that you can also use the colon symbol (:) to select a range of consecutive observations. For example, to remove observation numbers 2 to 10, use the command:

```
> dataset<-dataset[-c(2:10),]
```

Removing Duplicate Observations

To remove duplicates observations from a dataset, use the unique function.

```
> dataset<-unique(dataset)
```

To save the duplicates to a separate dataset before removing them, use the duplicated function:

```
> dups<-dataset[duplicated(dataset),]
```

Selecting a Subset of the Data

Observations can be selected according to selection criteria based on properties of the data, or randomly to form a random sample.

Selecting a Subset According to Selection Criteria

Sometimes you may need to select a subset of a dataset containing only those observations that match certain criteria, such as belonging to a particular category or where the value of one of the numeric variables falls within a given range. You can do this with the subset function. The command takes the general form:

```
> subset(dataset, condition)
```

For example, to select all of the observations from the people dataset where the value of the Eye.Color variable is Brown, use the command:

```
> subset(people, Eye.Color=="Brown")
```

	Subject	Eye.Color	Height	Hand.Span	Sex	Handedness
1	1	Brown	186	210	1	R
3	3	Brown	147	167	2	
5	5	Brown	170	193	1	R
11	11	Brown	163	223	1	R
16	16	Brown	173	196	1	R

Notice that you must use two equals signs rather than one.

To save the selected observations to a new dataset, assign the output to a new dataset name:

```
> browneyes<-subset(people, Eye.Color=="Brown")
```

To select all the observations for which a variable takes any one of a list of values, use the %in% operator. For example, to select all observations where Eye.Color is either Brown or Green, use the command:

```
> subset(people, Eye.Color %in% c("Brown", "Green"))
```

	Subject	Eye.Color	Height	Hand.Span	Sex	Handedness
1	1	Brown	186	210	1	R
2	2	Green	182	220	1	R
3	3	Brown	147	167	2	
4	4	Green	157	180	2	L
5	5	Brown	170	193	1	R
11	11	Brown	163	223	1	R
15	15	Green	160	190	2	
16	16	Brown	173	196	1	R

To select observations to exclude instead of to include, replace == with != (which mean "not equal to"). For example, to exclude all observations where the value of Eye.Color is equal to "Blue", use the command:

```
> subset(people, Eye.Color!="Blue")
```

Observations can also be selected according to the value of a numeric variable. For example, to select all observations from the people dataset where the Height variable is equal to 169, use the command:

```
> subset(people, Height==169)
```

	Subject	Eye.Color	Height	Hand.Span	Sex	Handedness
6	6	Blue	169	190	2	L

Notice that quotation marks are not required for numeric values.

With numeric variables, you can also use *relational operators* to select observations. For example, to select all observations for which the value of the Height variable is less than 165, use the command:

```
> subset(people, Height<165)
```

	Subject	Eye.Color	Height	Hand.Span	Sex	Handedness
3	3	Brown	147	167	2	
4	4	Green	157	180	2	L
11	11	Brown	163	223	1	R
15	15	Green	160	190	2	

Other relational operators you could use are > (greater than), >= (greater than or equal to) and <= (less than or equal to).

You can combine two or more conditions using the *AND* operator (denoted &) and the *OR* operator (denoted |). When two criteria are joined with the AND operator, R selects only those observations that meet both conditions. When they are joined with the OR operator, R selects the observations that meet either one of the conditions, or both.

For example, to select observations where Eye.Color is Brown *and* Height is less than 165, use the command:

```
> subset(people, Eye.Color=="Brown" & Height<165)
```

	Subject	Eye.Color	Height	Hand.Span	Sex	Handedness
3	3	Brown	147	167	2	
11	11	Brown	163	223	1	R

As well as selecting a subset of observations from the dataset, you can also use the select argument to select which variables to keep.

```
> subset(people, Height<165, select=c(Hand.Span, Height))
```

	Hand.Span	Height
3	167	147
4	180	157
11	223	163
15	190	160

Another way to subset a dataset is using bracket notation. For example, this command selects only those people with brown eyes:

```
> people[people$Eye.Color=="Brown",]
```

Note that it is not always necessary to subset a dataset before performing analysis. Many analysis functions have a subset argument within the function. This allows you to perform the analysis for a subset of the data. For example, this command creates a scatter plot of height against hand span, showing only males (i.e., where Sex is equal to 2):

```
> plot(Height~Hand.Span, people, subset=Sex==2)
```

You will learn more about scatter plots in Chapter 8.

Selecting a Random Sample from a Dataset

To select a random sample of observations from a dataset, use the sample function. For example, the following command selects a random sample of 50 observations from a dataset named dataset and saves them to new dataset named sampledata:

```
> sampledata<-dataset[sample(1:nrow(dataset), 50),]
```

By default, the sample function samples without replacement, so that no observation can be selected more than once. For this reason, the sample size must be less than the number of observations in the dataset. To sample *with* replacement, set the replace argument to T:

```
> sampledata<-dataset[sample(1:nrow(dataset), 50, replace=T),]
```

Sorting a Dataset

You can use the order function to sort a dataset. For example, to sort the people dataset by the Hand.Span variable, use the command:

```
> people<-people[order(people$Hand.Span),]
```

	Subject	Eye.Color	Height	Hand.Span	Sex	Handedness
3	3	Brown	147	167	2	
10	10	Blue	166	178	2	R
4	4	Green	157	180	2	L
6	6	Blue	169	190	2	L
15	15	Green	160	190	2	
5	5	Brown	170	193	1	R
9	9	Blue	166	193	2	R
16	16	Brown	173	196	1	R
1	1	Brown	186	210	1	R
8	8	Blue	173	211	1	R
13	13	Blue	176	214	1	
7	7	Brown	174	217	1	R
14	14	Blue	183	218	1	R
2	2	Green	182	220	1	R
11	11	Brown	163	223	1	R
12	12	Blue	184	225	1	R

To sort in decreasing instead of ascending order, set the decreasing argument to T:

```
> people<-people[order(people$Hand.Span, decreasing=T),]
```

You can also sort by more than one variable. To sort the dataset first by Sex and then by Height, use the command:

```
> people<-people[order(people$Sex, people$Height),]
```

Summary

You should now be able to prepare your dataset by creating any new variables required for your analysis, removing irrelevant data, and tidying the final dataset.

This table shows the most important commands covered in this chapter.

Task	Command
Rename variable	names(*dataset*)[n]<-"*Newname*"
View variable class	class(*dataset$variable*)
Change variable class to numeric	*dataset$var1*<-as.numeric(*dataset$var1*)
Change variable class to factor	*dataset$var1*<-as.factor(*dataset$var1*)
Change variable class to character	*dataset$var1*<-as.character(*dataset$var1*)

(continued)

Task	Command
Change variable class to date	*dataset$var1*<-as.Date(*dataset$var1*, "*format*")
Copy variable	*dataset$var2*<-*dataset$var1*
Divide variable into categories	*dataset$factor1*<-cut(*dataset$var1*, c(*1,2,3,4*), c("*Name1*", "*Name2*","*Name3*"))
Rename factor level	*dataset$variable*)[n]<-"*Newname*"
Reorder factor levels	*dataset$variable*, "*Level1*")
Join two character strings	*dataset$var3*<-paste(*dataset$var1*, *dataset$var2*)
Extract a substring	*dataset$var2*<-substring(*dataset$var1*, *first*, *last*)
Search character variable	grep("*search term*", *dataset$variable*)
Remove cases	*dataset*<-*dataset*[-c(*2,4,7*),]
Remove duplicates	*dataset*<-unique(*dataset*)
Select subset	subset(*dataset*, *variable*=="*value*")
Select random sample	*newdataset*<-*dataset*[sample(1:nrow(*dataset*), *samplesize*),]
Sort dataset	*dataset*<-*dataset*[order(*dataset$variable*),]

In Chapter 4, you will learn how to make more complex changes to datasets, including combining two or more datasets and changing the structure of the data.

■ ■ ■

Combining and Restructuring Datasets

This chapter explains how to make major changes to the structure of a dataset. You may need to combine data from two or more different sources into one dataset. This may be as simple as attaching one dataset to the bottom or side of another (known as *appending* or *concatenating*), or it may involving using a common variable to match the observations (known as merging). Alternatively, you may need to combine the values from several variables into one variable (stacking) or split one variable into several variables (unstacking). Finally, you may need to reshape a dataset so that data that was once held in separate variables in now held in separate observations, or vice versa (also known as *rotating* or *transposing*).

You will learn how to:

- append rows
- append columns
- merge two datasets using a common variable
- stack a dataset
- unstack a dataset
- reshape a dataset

This chapter uses the CIAdata1, CIAdata2, WHOdata, CPIdata, bigcats, endangered, grades1, resistance, and vitalsigns datasets, which are available with the downloads for this book (www.apress.com/9781484201404). For more details about these datasets, see Appendix C. It also uses the iris dataset, which is included with R. To view more information about this dataset, enter help(iris).

Appending Rows

The rbind function allows you to attach one dataset on to the bottom of the other, which is known as appending or concatenating the datasets. This is useful when you want to combine two datasets that contain different observations for the same variables, such as the CIAdata1 and CIAdata2 datasets shown in Figure 4-1a, b.

	country	lifeExp	urban	pcGDP
1	Finland	79.41	85	36700
2	Slovakia	76.03	55	23600
3	UK	80.17	80	36600
4	Ukraine	68.74	69	7300
5	Spain	81.27	77	31000

(a) CIAdata1

	country	pcGDP	lifeExp	urban
1	Italy	30900	81.86	68
2	Croatia	18400	75.99	58
3	Slovakia	23600	76.03	55

(b) CIAdata2

Figure 4-1. *Data from the CIA World Factbook (see Appendix C for more details)*

Before using the rbind function, make sure that each dataset contains the same number of variables and that all of the variable names match. You may need to remove or rename some variables first, as explained in the first two sections in Chapter 3. The variables do not need to be arranged in the same order within the datasets, as the rbind function automatically matches them by name.

Once the datasets are prepared, append them with the rbind function, as shown here for the CIAdata1 and CIAdata2 datasets:

```
> CIAdata<-rbind(CIAdata1, CIAdata2)
```

The new CIAdata dataset is shown in Figure 4-2. Notice that the new dataset contains all of the original data in the original order, including two copies of the data for Slovakia. The rbind function does not identify duplicates or sort the data. You can do this with the unique and order functions as explained in Chapter 3 (under "Removing Duplicate Observations" and "Sorting a Dataset").

	country	lifeExp	urban	pcGDP
1	Finland	79.41	85	36700
2	Slovakia	76.03	55	23600
3	UK	80.17	80	36600
4	Ukraine	68.74	69	7300
5	Spain	81.27	77	31000

(a) CIAdata1

	country	pcGDP	lifeExp	urban
1	Italy	30900	81.86	68
2	Croatia	18400	75.99	58
3	Slovakia	23600	76.03	55

(b) CIAdata2

	country	lifeExp	urban	pcGDP
1	Finland	79.41	85	36700
2	Slovakia	76.03	55	23600
3	UK	80.17	80	36600
4	Ukraine	68.74	69	7300
5	Spain	81.27	77	31000
6	Italy	81.86	68	30900
7	Croatia	75.99	58	18400
8	Slovakia	76.03	55	23600

(c) CIAdata

Figure 4-2. *The result of appending the CIAdata1 and CIAdata2 datasets to create the CIAdata dataset*

You can append three or more datasets in a similar manner:

```
> newdataset<-rbind(dataset1, dataset2, dataset3)
```

Appending Columns

The cbind function pastes one dataset on to the side of another. This is useful if the data from corresponding rows of each dataset belong to the same observation, as is the case for the CIAdata1 and WHOdata datasets shown in Figure 4-3a, b.

	country	lifeExp	urban	pcGDP
1	Finland	79.41	85	36700
2	Slovakia	76.03	55	23600
3	UK	80.17	80	36600
4	Ukraine	68.74	69	7300
5	Spain	81.27	77	31000

(a) CIAdata1

	alcohol	mortality
1	12.52	91
2	13.33	130
3	13.37	77
4	15.6	274
5	11.62	68

(b) WHOdata

Figure 4-3. *Data from the CIA World Factbook and from the World Health Organisation (see Appendix C for more details)*

You can only use the cbind function to combine datasets that have the same number of rows. If your datasets have a common variable or variables that can be used to match the observations, use the merge function to combine the datasets, as explained in the following section.

This command combines the CIAdata1 and WHOdata datasets to create a new dataset called CIAWHOdata; the result is shown in Figure 4-4c:

```
> CIAWHOdata<-cbind(CIAdata1, WHOdata)
```

	country	lifeExp	urban	pcGDP
1	Finland	79.41	85	36700
2	Slovakia	76.03	55	23600
3	UK	80.17	80	36600
4	Ukraine	68.74	69	7300
5	Spain	81.27	77	31000

(a) CIAdata1

	alcohol	mortality
1	12.52	91
2	13.33	130
3	13.37	77
4	15.6	274
5	11.62	68

(b) WHOdata

	country	lifeExp	urban	pcGDP	alcohol	mortality
1	Finland	79.41	85	36700	12.52	91
2	Slovakia	76.03	55	23600	13.33	130
3	UK	80.17	80	36600	13.37	77
4	Ukraine	68.74	69	7300	15.6	274
5	Spain	81.27	77	31000	11.62	68

(c) CIAWHOdata

Figure 4-4. *The result of appending the CIAdata1 and WHOdata datasets to create the CIAWHOdata dataset*

You can combine three or more datasets in a similar way:

```
> newdataset<-cbind(dataset1, dataset2, dataset3)
```

Merging Datasets by Common Variables

The merge function allows you to combine two datasets by matching the observations according to the values of common variables.

Consider the CIAdata1 and CPIdata datasets shown in Figure 4-5. The datasets have a common variable called country, which can be used to match corresponding observations.

	country	lifeExp	urban	pcGDP
1	Finland	79.41	85	36700
2	Slovakia	76.03	55	23600
3	UK	80.17	80	36600
4	Ukraine	68.74	69	7300
5	Spain	81.27	77	31000

(a) CIAdata1

	country	CPI
1	Spain	80.24
2	UK	100.13
3	Croatia	67.54
4	Italy	94.82
5	Ukraine	51.1
6	Finland	99.69
7	Spain	80.24

(b) CPIdata1

Figure 4-5. *Data from the CIA World Factbook and Numbeo (see Appendix C for more details)*

The following command shows how you would combine the CIAdata1 and CPIdata datasets; the result is shown in Figure 4-6:

```
> CIACPIdata<-merge(CIAdata1, CPIdata)
```

	country	lifeExp	urban	pcGDP
1	Finland	79.41	85	36700
2	Slovakia	76.03	55	23600
3	UK	80.17	80	36600
4	Ukraine	68.74	69	7300
5	Spain	81.27	77	31000

(a) CIAdata1

	country	CPI
1	Spain	80.24
2	UK	100.13
3	Croatia	67.54
4	Italy	94.82
5	Ukraine	51.1
6	Finland	99.69
7	Spain	80.24

(b) CPIdata

	country	lifeExp	urban	pcGDP	CPI
1	Finland	79.41	85	36700	99.69
2	Spain	81.27	77	31000	80.24
3	Spain	81.27	77	31000	80.24
4	UK	80.17	80	36600	100.13
5	Ukraine	68.74	69	7300	51.1

(c) CIACPIdata

Figure 4-6. *The result of merging the CIAdata1 and CPIdata datasets to create the CIACPIdata dataset*

The merge function identifies variables with the same name and uses them to match up the observations. In this example, both datasets contain a variable named country, so R automatically uses this variable to match the observations. If your datasets have more than one common variable, R matches the observations by the unique combinations of all of the common variables.

If the names of the common variables are not identical in the two datasets, the simplest solution is to rename the variables, as explained in Chapter 3 (see "Renaming Variables"). Alternatively, you can use the by.x and by.y arguments to specify which variables to use for matching the observations. For example, to merge two datasets in which the common variable is named var1 in the first dataset and VAR1 in the second dataset, use the command:

```
newdataset<-merge(dataset1, dataset2, by.x="var1", by.y="VAR1")
```

When you combine two datasets with the merge function, R automatically excludes any unmatched observations that appear in only one of the datasets. In Figure 4-6, you can see that Slovakia (which appears only in CIAdata1) and Italy and Croatia (which appear only in CPIdata) are all excluded from the merged dataset.

The all, all.x, and all.y arguments allow you to control how R deals with any unmatched observations. To keep all unmatched observations, set the all argument to T:

```
> allCIACPIdata<-merge(CIAdata1, CPIdata, all=T)
```

The results are shown in Figure 4-7. Notice that Slovakia, Italy, and Croatia have now been included, and where corresponding data is missing, the missing data symbol NA has been substituted.

	country	CPI
1	Spain	80.24
2	UK	100.13
3	Croatia	67.54
4	Italy	94.82
5	Ukraine	51.1
6	Finland	99.69
7	Spain	80.24

	country	lifeExp	urban	pcGDP
1	Finland	79.41	85	36700
2	Slovakia	76.03	55	23600
3	UK	80.17	80	36600
4	Ukraine	68.74	69	7300
5	Spain	81.27	77	31000

(a) CIAdata1   (b) CIAdata

	country	lifeExp	urban	pcGDP	CPI
1	Finland	79.41	85	36700	99.69
2	Slovakia	76.03	55	23600	NA
3	Spain	81.27	77	31000	80.24
4	Spain	81.27	77	31000	80.24
5	UK	80.17	80	36600	100.13
6	Ukraine	68.74	69	7300	51.1
7	Croatia	NA	NA	NA	67.54
8	Italy	NA	NA	NA	94.82

(c) allCIACPIdata

Figure 4-7. *The result of merging the CIAdata and CPIdata datasets with all=T to create the allCIACPIdata dataset*

Alternatively, you can use the all.x and all.y arguments to include unmatched cases from just one of the datasets instead of both. So the command:

```
> allCIAGPIdata<-merge(CIAdata1, CPIdata, all.x=T)
```

includes the data for Slovakia but not Italy and Croatia, while the command:

```
> allCIAGPIdata<-merge(CIAdata1, CPIdata, all.y=T)
```

includes the data for Italy and Croatia but not Slovakia.

When there are multiple matches for an observation, R creates an observation in the final dataset for every possible match. This is known as the *Cartesian product*. Consider the bigcats and endangered datasets shown in Figure 4-8a and b. Both datasets have two observations with a name of 'Leopard.' When these two datasets are merged by the Name variable, the final dataset (shown in Figure 4-8c) contains four rows for the Leopard data.

	Name	Weight
1	Tiger	203
2	Leopard	64
3	Leopard	58
4	Lion	200

(a) bigcats

	Name	Status
1	Lion	Vulnerable
2	Leopard	Least Concern
3	Tiger	Endangered
4	Leopard	Endangered

(b) endangered

	Name	Weight	Status
1	Leopard	64	Least Concern
2	Leopard	64	Endangered
3	Leopard	58	Least Concern
4	Leopard	58	Endangered
5	Lion	200	Vulnerable
6	Tiger	203	Endangered

(c) Merged dataset

Figure 4-8. *Datasets giving information about four big cat species (see Appendix C for more details)*

To ensure that you get the desired result from a merge, make sure that you are using a common variable or variables that allow R to correctly identify matches. You may also need to remove duplicate observations beforehand, as described in Chapter 3 (see "Removing Duplicate Observations").

R automatically sorts the merged dataset by the same variables that were used to match the observations. If you want to prevent this, set the sort argument to F:

```
> newdataset<-merge(dataset1, dataset2, sort=F)
```

Stacking and Unstacking a Dataset

Stacking a dataset means to convert it from *unstacked* form to *stacked* form, while unstacking a dataset is the opposite procedure, that is, converting a dataset from stacked form to unstacked form. To illustrate the difference between the two forms, consider the grades1 and grades2 datasets shown in Figure 4-9. The grades1 dataset is in unstacked form. It gives the test results of fifteen students, arranged in separate columns according to which class they belong. The grades2 dataset gives the same data in stacked form. Here the data is arranged in two columns, the first giving the test result and the second identifying to which class the student belongs.

	values	ind
1	87	ClassA
2	64	ClassA
3	89	ClassA
4	82	ClassA
5	59	ClassA
6	83	ClassB
7	97	ClassB
8	97	ClassB
9	99	ClassB
10	92	ClassB
11	80	ClassC
12	86	ClassC
13	80	ClassC
14	80	ClassC
15	84	ClassC

	ClassA	ClassB	ClassC
1	87	83	80
2	64	97	86
3	89	97	80
4	82	99	80
5	59	92	84

(a) grades1   (b) grades2

Figure 4-9. The same data in stacked and unstacked form

Having your data in stacked form is useful if you want to summarize all of the data values together, or if you want to model the values against the grouping variable. Having your data in unstacked form is useful as an alternative to dividing your data into subsets when you want to analyze the values in each group separately.

Many commonly used R functions such as aov (for analysis of variance), lm (for regression and general linear models), and glm (for generalized linear models) require data in stacked form. Most of R's hypothesis testing functions such as t.test are able to handle data in either stacked or unstacked form.

You should only stack or unstack a dataset if the values are numeric, in no particular order, and if there is no correspondence between the values in each of the groups. For more complex situations, such as when there is an additional variable present that indicates which of the values in each group correspond to each other, refer to "Reshaping a dataset."

Stacking Data

To convert a dataset from unstacked to stacked form, use the stack function:

```
> grades2<-stack(grades1)
```

To stack only some of the columns in your dataset, use the select argument. For example, to stack only the ClassA and ClassC variables from the grades1 dataset, use the command:

```
> newdataset<-stack(grades1, select=c("ClassA", "ClassC"))
```

If there are any non-numeric variables in your dataset, then R will exclude them from the stacked dataset.

The stack function automatically names the new variables values and ind, but you can change the names to something more informative with the names function:

```
> names(grades2)<-c("Result", "Class")
```

Figure 4-10 shows the final stacked dataset.

	Result	Class
1	87	ClassA
2	64	ClassA
3	89	ClassA
4	82	ClassA
5	59	ClassA
6	83	ClassB
7	97	ClassB
8	97	ClassB
9	99	ClassB
10	92	ClassB
11	80	ClassC
12	86	ClassC
13	80	ClassC
14	80	ClassC
15	84	ClassC

Figure 4-10. *The* grades2 *dataset, after the variable names are changed*

Unstacking Data

You can unstack a dataset with the unstack function. If your dataset has only two variables, the first of which gives the data values and the second of which identifies which group the observation belongs to (like the grades2 dataset), use the function:

```
> grades1<-unstack(grades2)
```

If your dataset has more than two variables or the variables are in reverse order then you must specify which variables you want to unstack:

```
> unstackeddata<-unstack(stackeddata, values~groups)
```

Here, values is the name of the variable containing the data values and groups is the name of the variable that identifies to which group the observation belongs.

For example, to unstack the Sepal.Width column in the iris dataset (included with R) to create a new dataset with a column for each iris species, use the command:

```
> sepalwidths<-unstack(iris, Sepal.Width~Species)
```

If you unstack a variable that does not have an equal number of values in each group, R cannot arrange the values to create a new data frame. In this case, R creates a *list* object instead of a data frame. However, you can still access each group of values in the new object by using the dollar notation:

```
> unstackeddata$groupA
```

Reshaping a Dataset

Reshaping a dataset is also known as rotating or transforming a dataset. It usually applies to datasets in which repeat measurements have been taken, but it is useful in other situations, too. Reshaping is the process of changing between long form (with repeat measurements in separate columns) and wide form (with repeat measurements in the same column).

To illustrate the difference between the long and wide forms, consider the resistance dataset shown in Figure 4-11a, which gives cubic resistance measurements for four concrete formulations taken at three, seven, and fourteen days after setting. This dataset is in wide form. Figure 4-11b shows the same dataset in long form, with all the measurements in one variable and another variable giving the day that the measurement was taken.

	Formula	Day3	Day7	Day14
1	A	10.8	18.5	41.3
2	B	20.1	29.2	37
3	C	3.7	17.5	28.9
4	D	18.1	27.2	37.6

(a) resistance

	row.names	Formula	Day	Resistance
1	A.3	A	3	10.8
2	B.3	B	3	20.1
3	C.3	C	3	3.7
4	D.3	D	3	18.1
5	A.7	A	7	18.5
6	B.7	B	7	29.2
7	C.7	C	7	17.5
8	D.7	D	7	27.2
9	A.14	A	14	41.3
10	B.14	B	14	37
11	C.14	C	14	28.9
12	D.14	D	14	37.6

(b) resistance2

Figure 4-11. *The resistance dataset in long and wide forms (see Appendix C for more details)*

Converting your dataset to long form is useful if you want to build a model that includes the time point as an explanatory variable. For example, converting the resistance dataset to long form allows you to model the resistance against the number of days elapsed since setting the concrete.

The reshape function allows you to convert a dataset from wide to long form. For example, to convert the resistance dataset to long form, use the command:

```
> resistance2<-reshape(resistance, direction="long", varying=list(c("Day3", "Day7", "Day14")),
                times=c(3, 7, 14), idvar="Formula", v.names="Resistance", timevar="Day")
```

Use the varying argument to specify the variables that you want to combine into one column. Use the times argument to give the new time values or replicate numbers for each of these variables. Note that the times can be character values instead of numeric. Use the idvar argument to specify the variable that will group the records together. The v.names and timevar arguments are optional and give the names for the new variables.

The new dataset is shown in Figure 4-11b.

You can also use the reshape function to perform the opposite procedure, that is, converting a dataset from long to wide form.

Consider the vitalsigns dataset shown in Figure 4-12a, which gives measurements of systolic blood pressure, diastolic blood pressure, and pulse for four patients. All of the measurements are held in the result variable, while the test variable identifies the parameter.

	subject	test	result
1	1733	Pulse	120
2	1733	DiaBP	79
3	1733	SysBP	79
4	1734	Pulse	121
5	1734	DiaBP	86
6	1734	SysBP	72
7	1735	Pulse	130
8	1735	DiaBP	94
9	1735	SysBP	74
10	1736	Pulse	142
11	1736	DiaBP	99
12	1736	SysBP	87

(a) vitalsigns

	subject	result.Pulse	result.DiaBP	result.SysBP
1	1733	120	79	79
2	1734	121	86	72
3	1735	130	94	74
4	1736	142	99	87

(b) vitalsigns2

Figure 4-12. *The* vitalsigns *dataset in long and wide forms (see Appendix C for more details)*

Converting your dataset to wide form is useful if you want to compare data from two or more time points or parameters, or include them as distinct variables in the same model. For example, reshaping the vitalsigns dataset allows you to calculate the correlation between systolic and diastolic blood pressure.

To split the different types of measurements in to separate columns, use the command:

```
> vitalsigns2<-reshape(vitalsigns, direction="wide", v.names="result", timevar="test",
                       idvar="subject")
```

Use the v.names argument to specify the variable that you want to separate into different columns. Use the timevar argument to specify the variable that indicates which column the value belongs to (i.e., the variable giving the time point, replicate number, or category for the record). Use the idvar argument to specify which variable is used to group the records together.

Figure 4-12b shows the new wide form dataset. Notice that R automatically generates names for the new variables. Alternatively, you can specify the new names with the varying argument:

```
> vitalsigns2<-reshape(vitalsigns, direction="wide", v.names="result", timevar="test",
                       idvar="subject", varying=list(c("SysBP", "DiaBP", "Pulse")))
```

Summary

You should now be able to combine two or more datasets, either by appending them or by merging them using a common variable. You should understand the difference between stacked and unstacked forms, and be able to stack and unstack a dataset as required. You should also understand the difference between long and wide forms and be able to convert between the two.

This table summarizes the most important commands covered in this chapter.

Task	Command
Append datasets vertically	rbind(*dataset1*, *dataset2*)
Append datasets horizontally	cbind(*dataset1*, *dataset2*)
Merge datasets	merge(*dataset1*, *dataset2*)
Stack dataset	stack(*dataset*, select=c("*var1*", "*var2*", "*var3*"))
Unstack dataset	unstack(*dataset*, *values~groups*)
Reshape (wide to long)	reshape(*dataset*, direction="long", varying=list(c ("*var1*", "*var2*", "*var3*")), times=c(*t1*,*t2*,*t3*), idvar="*identifier*")
Reshape (long to wide)	reshape(*dataset*, direction="wide", v.names="*values*", timevar="*groups*", idvar="*identifier*")

In the next chapter, we will move on to summarizing your data, beginning with continuous variables.

CHAPTER 5

■ ■ ■

Summary Statistics for Continuous Variables

A good place to begin analyzing your data is with some simple summary statistics. This chapter explains how to calculate summary statistics for continuous variables.

Two main types of summary statistics are univariate summary statistics and measures of association. Univariate statistics are those that are calculated from a single variable. This includes measures of location such as the mean and median, and measures of dispersion such as the variance, standard deviation and range. Measures of association summarise the relationship between two variables, and include the covariance, Pearson's correlation, and Spearman's correlation.

Also covered in this chapter are methods that help you to make inferences about the population from which a sample is drawn. These include the the Shapiro-Wilk and Kolmogorov-Smirnov tests, and confidence and prediction intervals.

You will learn how to:

- calculate univariate statistics

- calculate statistics for different groups of observations

- calculate the covariance, Pearson's correlation, and Spearman's correlation between two variables

- perform a hypothesis test to check whether a correlation is statistically significant

- perform the Shapiro-Wilk and Kolmogorov-Smirnov tests

- calculate confidence and prediction intervals

This chapter uses the `trees`, `iris`, `warpbreaks`, and `PlantGrowth` datasets, which are included with R, and the `bottles` dataset, which is available with the downloads for this book. It is helpful to become familiar with them before beginning the chapter. For the datasets included with R, you can view additional information about them by entering `help(`*datasetname*`)`. For more information about the `bottles` dataset, see Appendix C.

Univariate Statistics

To produce a summary of all the variables in a dataset, use the `summary` function. The function summarizes each variable in a manner suitable for its class. For numeric variables, it gives the mean, median, range, and interquartile range. For factor variables, it gives the number in each category. If a variable has any missing values, it will tell you how many.

```
> summary(iris)
```

```
 Sepal.Length    Sepal.Width    Petal.Length    Petal.Width          Species
 Min.   :4.300   Min.   :2.000  Min.   :1.000   Min.   :0.100   setosa    :50
 1st Qu.:5.100   1st Qu.:2.800  1st Qu.:1.600   1st Qu.:0.300   versicolor:50
 Median :5.800   Median :3.000  Median :4.350   Median :1.300   virginica :50
 Mean   :5.843   Mean   :3.057  Mean   :3.758   Mean   :1.199
 3rd Qu.:6.400   3rd Qu.:3.300  3rd Qu.:5.100   3rd Qu.:1.800
 Max.   :7.900   Max.   :4.400  Max.   :6.900   Max.   :2.500
```

To calculate a particular statistic for a single variable, use the relevant function from Table 5-1.

Table 5-1. *Functions for Summarizing Continuous Variables;*
Those Marked with an Asterisk Give a Single Value as Output

Statistic	Function
Mean*	mean
Median*	median
Standard deviation*	sd
Median absolute deviate*	mad
Variance*	var
Maximum value*	max
Minimum value*	min
Interquartile range*	IQR
Range	range
Quantiles	quantile
Tukey five-number summary	fivenum
Sum*	sum
Product*	prod
Number of observations*	length

For example, to calculate the mean tree height, use the command:

```
> mean(trees$Height)
```

```
[1] 76
```

If the variable has any missing data values, set the na.rm argument to T as shown below. This tells R to ignore any missing values when calculating the statistic. Otherwise, the result will be either another missing value or an error message, depending on the function:

```
> mean(dataset$variable, na.rm=T)
```

To calculate a particular statistic for each of the variables in a dataset simultaneously, use the sapply function with any of the statistics in Table 5-1:

```
> sapply(trees, mean)
```

```
   Girth    Height    Volume
13.24839 76.00000 30.17097
```

Again, if the dataset has any missing values then set the na.rm argument to T:

```
> sapply(dataset, mean, na.rm=T)
```

If any of the variables in your dataset are nonnumeric, the sapply function behaves inconsistently. For example, this command attempts to calculate the maximum value for each of the variables in the iris dataset. R returns an error message because the fifth variable in the dataset is a factor variable:

```
> sapply(iris, max)
```

```
Error in Summary.factor(c(1L, 1L, 1L, 1L, 1L, 1L, 1L, 1L, 1L, 1L, 1L,  :
max not meaningful for factors
```

To avoid this problem, exclude any nonnumeric variables from the dataset by using bracket notation or the subset function, as described in Chapters 1 (under "Data Frames") and 3 (under "Selecting a Subset of the Data"):

```
> sapply(iris[-5], max)
```

```
Sepal.Length  Sepal.Width Petal.Length  Petal.Width
         7.9          4.4          6.9          2.5
```

Statistics by Group

You may want to group the values of a numeric variable according to the levels of a factor and calculate a statistic for each group. There are two functions that allow you to do this, called tapply and aggregate.

You can use the tapply function with any of the statistics in Table 5-1. For example, to calculate the mean sepal width for each species for the iris dataset:

```
> tapply(iris$Sepal.Width, iris$Species, mean)
```

```
    setosa versicolor  virginica
     3.428      2.770      2.974
```

If the numeric variable has any missing data, set the na.rm argument to T.

```
> tapply(dataset$variable, dataset$factor1, mean, na.rm=T)
```

You can also group the data by more than one factor, by nesting the factor variables inside the `list` function. R calculates the statistic separately for each combination of factor levels. For example, to display the median number of breaks for each combination of tension and wool type for the `warpbreaks` dataset:

```
> tapply(warpbreaks$breaks, list(warpbreaks$wool, warpbreaks$tension), median)
```

```
   L  M  H
A 51 21 24
B 29 28 17
```

When using more than one grouping variable, you can only use statistical functions that give a single value as output (i.e., those marked with an asterisk in Table 5-1).

Alternatively, you can also use the `aggregate` function to summarize variables by groups. Using the `aggregate` function has the advantage that you can summarize several continuous variables simultaneously. It can also be used with statistical functions that give more than one value as output (such as `range` and `quantile`). However, the results are displayed a little differently, so it is a matter of personal preference whether to use `tapply` or `aggregate`.

To calculate the mean sepal width for each species, use the `aggregate` function as shown:

```
> aggregate(Sepal.Width~Species, iris, mean)
```

```
     Species Sepal.Width
1      setosa       3.428
2 versicolor       2.770
3  virginica       2.974
```

Again, you can also use more than one grouping variable. For example, to calculate the median number of breaks for each combination of wool and tension for the `warpbreaks` dataset:

```
> aggregate(breaks~wool+tension, warpbreaks, median)
```

```
  wool tension breaks
1    A       L     51
2    B       L     29
3    A       M     21
4    B       M     28
5    A       H     24
6    B       H     17
```

To summarize two or more continuous variables simultaneously, nest them inside the `cbind` function as shown:

```
> aggregate(cbind(Sepal.Width, Sepal.Length)~Species, iris, mean)
```

```
     Species Sepal.Width Sepal.Length
1      setosa       3.428        5.006
2 versicolor       2.770        5.936
3  virginica       2.974        6.588
```

You can save the output to a new data frame, as shown here. This allows you to use the results for further analysis:

```
> sepalmeans<-aggregate(cbind(Sepal.Width, Sepal.Length)~Species, iris, mean)
```

Remember to set the na.rm argument to T if any of the continuous variables have missing values.

Measures of Association

The *association* between two variables is a relationship between them, such that if you know the value of one variable, it tells you something about the value of the other. Positive association means that as the value of one variable increases, the value of the other also tends to increase. Negative association means that as the value of one variable increases, the value of the other tends to decrease. The most commonly used measures of association are:

Covariance: A measure of the linear association between two continuous variables. Covariance is scale dependent, meaning that the value depends on the units of measurements used for the variables. For this reason, it is difficult to directly interpret the covariance value. The higher the absolute covariance between two variables, the greater the association. Positive values indicate positive association and negative values indicate negative association.

Pearson's correlation coefficient (denoted *r*): A scale independent measure of association, meaning that the value is not affected by the unit of measurement. The correlation can take values between -1 and 1, where -1 indicates perfect negative correlation, 0 indicates no correlation and 1 indicates perfect positive correlation. The correlation coefficient only measures *linear* relationships, so it is important to check for nonlinear relationships with a scatter plot (see the "Scatter Plots" section in Chapter 8).

Spearman's rank correlation coefficient: A nonparametric alternative to the Pearson's correlation coefficient, which measures nonlinear as well as linear relationships. It also takes values between -1 (perfect negative correlation) and 1 (perfect positive correlation), with a value of 0 indicating no correlation. Spearman's correlation can be calculated for ranked as well as continuous data.

The following subsections explain how to calculate each of these measures of association in R.

Covariance

To calculate the covariance between two variables, use the cov function:

```
> cov(trees$Height, trees$Volume)
```

```
[1] 62.66
```

You can also create a covariance matrix for a whole dataset, which shows the covariance for each pair of variables:

```
> cov(trees)
```

	Girth	Height	Volume
Girth	9.847914	10.38333	49.88812
Height	10.383333	40.60000	62.66000
Volume	49.888118	62.66000	270.20280

From the output, you can see that the covariance between tree girth and tree height is 10.38. The values along the diagonal of the matrix give the variance of the variables. For example, the tree volumes have a variance of 270.2.

If your dataset has any nonnumeric variables, R will display an error message:

```
> cov(iris)
```

```
Error: is.numeric(x) || is.logical(x) is not TRUE
```

To avoid this problem, exclude the nonnumeric variables using bracket notation or the subset function:

```
> cov(iris[-5])
```

	Sepal.Length	Sepal.Width	Petal.Length	Petal.Width
Sepal.Length	0.6856935	-0.0424340	1.2743154	0.5162707
Sepal.Width	-0.0424340	0.1899794	-0.3296564	-0.1216394
Petal.Length	1.2743154	-0.3296564	3.1162779	1.2956094
Petal.Width	0.5162707	-0.1216394	1.2956094	0.5810063

If any of the variables have missing values, set the use argument to "pairwise" as shown here. Otherwise, the covariance matrix will also have missing values:

```
> cov(dataset, use="pairwise")
```

Another useful option for the use argument is "complete". This option completely excludes observations that have missing values for any of the variables, while "pairwise" excludes only those observations that have missing values for either of the variables in a given pair. Enter help(cov) to view more details about these options.

Pearson's Correlation Coefficient

To calculate the Pearson's correlation coefficient between two variables, use the cor function:

```
> cor(trees$Girth, trees$Volume)
```

```
[1] 0.9671194
```

The value is very close to 1, which indicates a very strong positive correlation between tree girth and tree volume. This means that trees with a larger girth tend to have a larger volume.

You can also create a correlation matrix for a whole dataset:

```
> cor(trees)
```

```
          Girth    Height    Volume
Girth   1.0000000 0.5192801 0.9671194
Height  0.5192801 1.0000000 0.5982497
Volume  0.9671194 0.5982497 1.0000000
```

Notice that the values along the diagonal of the matrix are all equal to 1, because a variable always correlates perfectly with itself.

Remember to exclude any nonnumeric variables using bracket notation or the subset function, as shown here for the iris dataset:

```
> cor(iris[-5])
```

Again if any of the variables have missing values, set the use argument to "pairwise":

```
> cor(dataset, use="pairwise")
```

Spearman's Rank Correlation Coefficient

The cor function can also calculate the Spearman's rank correlation coefficient between two variables. Set the method argument to "spearman":

```
> cor(trees$Girth, trees$Volume, method="spearman")
```

```
[1] 0.9547151
```

You can also create a correlation matrix for a whole dataset. Remember to exclude any nonnumeric variables using bracket notation:

```
> cor(trees, method="spearman")
```

```
          Girth    Height    Volume
Girth   1.0000000 0.4408387 0.9547151
Height  0.4408387 1.0000000 0.5787101
Volume  0.9547151 0.5787101 1.0000000
```

If any of the variables have missing values, set the use argument to "pairwise":

```
> cor(dataset$var1, dataset$var2, method="spearman", use="pairwise")
```

Hypothesis Test of Correlation

A hypothesis test of correlation determines whether a correlation is statistically significant. The null hypothesis for the test is that the population correlation is equal to zero, meaning that there is no correlation between the variables. The alternative hypothesis is that the population correlation is not equal to zero, meaning that there is some correlation between the variables. You can also perform a one-sided test, where the alternative hypothesis is either that the population correlation is greater than zero (the variables are positively correlated) or that the population correlation is less than zero (the variables are negatively correlated).

■ **Note** See Chapter 10 for more details about hypothesis testing.

You can perform a test of the correlation between two variables with the cor.test function:

```
> cor.test(dataset$var1, dataset$var2)
```

By default, R performs a test of the Pearson's correlation. If you would prefer to test the Spearman's correlation, set the method argument to "spearman":

```
> cor.test(dataset$var1, dataset$var2, method="spearman")
```

By default, R performs a two-sided test, but you can adjust this by setting the alternative argument to "less" or "greater" as required:

```
> cor.test(dataset$var1, dataset$var2, alternative="greater")
```

The output includes a 95% confidence interval for the correlation estimate. To adjust the size of this interval, use the conf.level argument:

```
> cor.test(dataset$var1, dataset$var2, conf.level=0.99)
```

EXAMPLE 5-1.
HYPOTHESIS TEST OF CORRELATION USING THE TREES DATASET

Suppose that you want to perform a hypothesis test to help determine whether the correlation between tree girth and tree volume is statistically significant.

To perform a two-sided test of the Pearson's product moment correlation between tree girth and volume at the 5% significance level, use the command:

```
> cor.test(trees$Girth, trees$Volume)
```

The output is shown here:

```
        Pearson's product-moment correlation

data:  trees$Girth and trees$Volume
t = 20.4783, df = 29, p-value < 2.2e-16
alternative hypothesis: true correlation is not equal to 0
95 percent confidence interval:
  0.9322519 0.9841887
sample estimates:
        cor
  0.9671194
```

The correlation is estimated at 0.967, with a 95% confidence interval of 0.932 to 0.984. This means that as tree girth increases, tree volume tends to increase also.

Because the p-value of 2.2e-16 is much less than the significance level of 0.05, we can reject the null hypothesis that there is no correlation between girth and volume, in favor of the alternative hypothesis that the two are correlated.

SCIENTIFIC NOTATION

Scientific notation is a way of expressing very large or very small numbers more compactly.

For example, the number 720000 is equal to 7.2×100000 or 7.2×10^5. R and many other programming languages use the letter e to express "times ten to the power of", so that 7.2×10^5 is displayed as 7.2e5.

Scientific notation works similarly for very small numbers, for example the number 0.000072 is equal to 7.2×0.00001 or 7.2×10^{-5}. Using the R notation, this would be displayed as 7.2e-5.

Comparing a Sample with a Specified Distribution

Sometimes you may wish to determine whether your sample is consistent with having been drawn from a particular type of distribution such as the normal distribution. This is useful because many statistical techniques (such as analysis of variance) are only suitable for normally distributed data.

Two methods that allow you to do this are the Shapiro-Wilk and Kolmogorov-Smirnov tests. To visually assess how well your data fits the normal distribution, use a histogram or normal probability plot (covered in Chapter 8).

Shapiro-Wilk Test

The Shapiro-Wilk test is a hypothesis test that can help to determine whether a sample has been drawn from a normal distribution. The null hypothesis for the test is that the sample is drawn from a normal distribution and the alternative hypothesis is that it is not.

You can perform a Shapiro-Wilk test with the shapiro.test function.

```
> shapiro.test(dataset$variable)
```

```
                          EXAMPLE 5-2.
          SHAPIRO-WILK TEST USING THE TREES DATASET
```

Suppose that you want to perform a Shapiro-Wilk test to help determine whether the tree heights follow a normal distribution. You will use a 5% significance level. To perform the test, use the command:

```
> shapiro.test(trees$Height)
```

```
                    Shapiro-Wilk normality test

data:  trees$Height
W = 0.9655, p-value = 0.4034
```

From the output we can see that the p-value for the test is 0.4034. Because this is not less than our significance level of 0.05, we cannot reject the null hypothesis. This means there is no evidence that the tree heights do not follow a normal distribution.

Kolmogorov-Smirnov Test

A one-sample Kolmogorov-Smirnov test helps to determine whether a sample is drawn from a particular theoretical distribution. It has the null hypothesis that the sample is drawn from the distribution and the alternative hypothesis that it is not.

A two-sample Kolmogorov-Smirnov test helps to determine whether two samples are drawn from the same distribution. It has the null hypothesis that they are drawn from the same distribution and the alternative hypothesis that they are not.

Note that both the one and two-sample Kolmogor-Smirnov tests require continuous data. The test cannot be performed if your data contains ties (i.e some of the values are exactly equal). This is may be an issue if your data is not recorded to a sufficient number of decimal places.

You can perform a Kolmogorov-Smirnov test with the ks.test function. To perform a one-sample test with the null hypothesis that the sample is drawn from a normal distribution with a mean of 100 and a standard deviation of 10, use the command:

```
> ks.test(dataset$variable, "pnorm", 100, 10)
```

To test a sample against another theoretical distribution, replace "pnorm" with the relevant cumulative distribution function. A list of functions for standard probability distributions is given in Table 7.1 in Chapter 7. You must also substitute the mean and standard deviation with any parameters relevant to the distribution. Use the help function to check the parameters for the distribution of interest.

To perform a two-sample test to determine whether two samples are drawn from the same distribution, use the command:

```
> ks.test(dataset$sample1, dataset$sample2)
```

If your data is in stacked form (with the values for both samples in one variable), you must first unstack the dataset as explained in Chapter 4 (under "Unstacking Data").

EXAMPLE 5-3.
ONE-SAMPLE KOLMOGOROV-SMIRNOV TEST USING BOTTLES DATA

Consider the bottles dataset, which is available with the downloads for this book. The dataset gives data for a sample of 20 bottles of soft drink taken from a filling line. The dataset contains one variable named Volume, which gives the volume of liquid in millilitres for each of the bottles.

The bottle filling volume is believed to follow a normal distribution with a mean of 500 milliliters and a standard deviation of 25 milliliters. Suppose that you wish to use a one-sample Kolmogorov-Smirnov test to determine whether the data is consistent with this theory. The test has the null hypothesis that the bottles volumes are drawn from the described distribution, and the alternative hypothesis that they are not. A significance level of 0.05 will be used for the test.

To perform the test, use the command:

```
> ks.test(bottles$Volume, "pnorm", 500, 25)
```

This gives the following output:

```
        One-sample Kolmogorov-Smirnov test

data:  bottles$Volume
D = 0.2288, p-value = 0.2108
alternative hypothesis: two-sided
```

From the output we can see that the p-value for the test is 0.2108. As this is not less than the significance level of 0.05, we cannot reject the null hypothesis. This means that there is no evidence that the bottle volumes are not drawn from the described normal distribution.

EXAMPLE 5-4.
TWO-SAMPLE KOLMOGOROV-SMIRNOV TEST USING
THE PLANTGROWTH DATA

Consider the PlantGrowth dataset (included with R), which gives the dried weight of thirty batches of plants, each of which received one of three different treatments. The weight variable gives the weight of the batch and the groups variable gives the treatment received (ctrl, trt1 or trt2).

Suppose that you want to use a two-sample Kolmogorov-Smirnov test to determine whether batch weight has the same distribution for the treatment groups trt1 and trt2.

First you need to unstack the data with the unstack function.

```
> PlantGrowth2<-unstack(PlantGrowth)
```

Once the data is in unstacked form, perform the test as shown:

```
> ks.test(PlantGrowth2$trt1, PlantGrowth2$trt2)
```

This gives the following output:

```
        Two-sample Kolmogorov-Smirnov test

data:  PlantGrowth2$trt1 and PlantGrowth2$trt2
D = 0.8, p-value = 0.002057
alternative hypothesis: two-sided
```

The p-value of 0.002057 is less than the significance level of 0.05. This means that there is evidence to reject the null hypothesis that the batch weight distribution is the same for both treatment groups, in favor of the alternative hypothesis that the batch weight distribution is different for the two treatment groups.

Confidence Intervals and Prediction Intervals

A confidence interval for the population mean gives an indication of how accurately the sample mean estimates the population mean. A 95% confidence interval is defined as an interval calculated in such a way that if a large number of samples were drawn from a population and the interval calculated for each of these samples, 95% of the intervals will contain the true population mean value.

A prediction interval gives an indication of how accurately the sample mean predicts the value of a further observation drawn from the population.

The simplest way to obtain a confidence interval for a sample mean is with the t.test function, which provides one with the output. The "Student's T-Tests" section in Chapter 10 discusses the function in more detail. If you are only interested in obtaining a confidence interval, use the command:

```
> t.test(trees$Height)
```

```
        One Sample t-test

data:  trees$Height
t = 66.4097, df = 30, p-value < 2.2e-16
alternative hypothesis: true mean is not equal to 0
95 percent confidence interval:
  73.6628 78.3372
sample estimates:
mean of x
      76
```

From the results, you can see that the mean tree height is 76 feet, with a 95% confidence interval of 73.7 to 78.3 feet.

By default, R calculates a 95% interval. For a different size confidence interval such as 99%, adjust the conf.level argument as shown:

```
> t.test(trees$Height, conf.level=0.99)
```

You can also calculate one-sided confidence intervals. For an upper confidence interval, set the alternative argument to "less". For a lower confidence interval, set it to "greater".

```
> t.test(dataset$variable, alternative="greater")
```

To calculate a prediction interval for the tree heights, use the command:

```
> predict(lm(trees$Height~1), interval="prediction")[1,]
```

```
    fit    lwr     upr
76.0000 62.7788 89.2212
Warning message:
In predict.lm(lm(trees$Height ~ 1), interval = "prediction") :
  predictions on current data refer to _future_ responses
```

From the output you can see the the prediction interval for the tree heights is 62.8 to 89.2 feet.

Again, you can adjust the confidence level with the level argument as shown:

```
> predict(lm(dataset$variable~1), interval="prediction", level=0.99)[1,]
```

The commands for creating prediction intervals use the lm and predict functions. You will learn more about these functions in Chapter 11, which will help you to understand what these complex commands are doing. For now, you can just use them by substituting the appropriate dataset and variable names.

Summary

You should now be able to use R to summarize the continuous variables in your dataset and examine the relationship between them. You should also be able to make some inferences about the population from which a sample is drawn, such as whether it has a normal distribution.

This table summarizes the most important commands covered in this chapter.

Task	Command
Statistic for each variable	sapply(*dataset*, *statistic*)
Statistic by group	tapply(*dataset$var1*, *dataset$factor1*, *statistic*)
Statistic by group	aggregate(*variable~factor*, *dataset*, *statistic*)
Covariance	cov(*dataset$var1*, *dataset$var2*)
Covariance matrix	cov(*dataset*)
Pearson's correlation coefficient	cor(*dataset$var1*, *dataset$var2*)
Correlation matrix	cor(*dataset*)
Spearman's rank correlation coefficient	cor(*dataset$var1*, *dataset$var2*, method="spearman")
Spearman's correlation matrix	cor(*dataset*, method="spearman")
Hypothesis test of correlation	cor.test(*dataset$var1*, *dataset$var2*)
Shapiro-Wilk test	shapiro.test(*dataset$variable*)
One-sample Kolmogorov-Smirnov test	ks.test(*dataset$sample1*, "pnorm", *mean, sd*)
Two-sample Kolmogorov-Smirnov test	ks.test(*dataset$sample1*, *dataset$sample2*)
Confidence interval	t.test(*dataset$variable*)
Prediction interval	predict(lm(*dataset$variable~1*), interval="prediction")[1,]

Now that you have learned how to summarize continuous variables, we can move on to the next chapter, which will look at categorical variables.

CHAPTER 6

■ ■ ■

Tabular Data

This chapter explains how to summarize and calculate statistics for categorical variables. Categorical data is normally summarized in a frequency table (for one variable) or contingency table (for two or more variables), which is a table showing the number of values in each category or in each combination of categories. Data summarized in this form is known as tabular data.

As well as summarizing your categorical data in a table, you may need to compare it with a hypothesized distribution using the chi-square goodness-of-fit test. You may also want to determine whether there is any association between two or more categorical variables. The chi-square test of association and the Fisher's exact test are two methods that can help you with this.

You will learn how to:

- create frequency and contingency tables to summarise categorical data, and how to store the results as a table object

- display your table with percentages or marginal sums

- perform a chi-square goodness-of-fit test

- perform a chi-square test of association to identify a relationship between two or more categorical variables

- perform Fisher's exact test of association for a two-by-two contingency table

- perform a test of proportions

This chapter uses the warpbreaks and esoph datasets and the Titanic table object, which are included with R. You can view more information about them by entering help(*datasetname*). It also uses the people2 dataset (which is a modified version of the people dataset introduced in Chapter 3), and the apartments dataset. These are both available with the downloads for this book and described in Appendix C.

Frequency Tables

Frequency tables summarize a categorical variable by displaying the number of observations belonging to each category. Frequency tables for two or more categorical variables (known as contingency tables or cross tabs) summarize the relationship between two or more categorical variables by displaying the number of observations that fall into each combination of categories.

In R, a table is also a type of object that holds tabulated data. There are some example table objects included with R, such as HairEyeColor and Titanic.

Creating Tables

To create a one-dimensional frequency table showing the number of observations for each level of a factor variable, use the table function:

```
> table(people2$Eye.Color)
```

```
Blue Brown Green
   7     6     3
```

Whereas R allows you to create tables from any type of variable, they are only really meaningful for factor variables with a relatively small number of values. If you want to include a continuous variable, first divide it into categories with the cut function, as explained in Chapter 3 under "Dividing a Continuous Variable into Categories." To add an additional column to the table showing the numbers of missing values (if there are any), set the useNA argument:

```
> table(dataset$factor1, useNA="ifany")
```

To create a two-dimensional contingency table, give two variables as input:

```
> table(people2$Eye.Color, people2$Sex)
```

	Male	Female
Blue	4	3
Brown	5	1
Green	1	2

Similarly, you can create contingency tables for three or more variables:

```
> table(people2$Eye.Color, people2$Height.Cat, people2$Sex)
```

```
, ,  = Male
```

	Medium	Short	Tall
Blue	2	0	2
Brown	4	0	1
Green	0	0	1

```
, ,  = Female
```

	Medium	Short	Tall
Blue	3	0	0
Brown	0	0	0
Green	0	2	0

To save a table as a table object, assign the output of the table function to a new object name:

```
> sexeyetable<-table(people2$Eye.Color, people2$Sex)
```

Once you have created a table object, you can apply further functions to the object to create output that is relevant to tabular data. For example, you could use the pie function to create a pie chart showing the proportion of people with each eye color, or use the summary function to test for association between eye color and sex. You will learn about some of these functions later in this chapter. Save the sexeyetable object, as it will be useful later on.

Displaying Tables

There are a few functions that allow you to present table objects in different ways. These include ftable, prop.table, and addmargins.

The ftable function displays your table in a more compact way, which is useful for tables with three or more dimensions:

```
> ftable(Titanic)
```

```
                    Survived  No Yes
Class Sex    Age
1st   Male   Child             0   5
             Adult           118  57
      Female Child             0   1
             Adult             4 140
2nd   Male   Child             0  11
             Adult           154  14
      Female Child             0  13
             Adult            13  80
3rd   Male   Child            35  13
             Adult           387  75
      Female Child            17  14
             Adult            89  76
Crew  Male   Child             0   0
             Adult           670 192
      Female Child             0   0
             Adult             3  20
```

The prop.table function displays the table with each cell count expressed as a proportion of the total count:

```
> prop.table(sexeyetable)
```

```
        Male Female
Blue  0.2500 0.1875
Brown 0.3125 0.0625
Green 0.0625 0.1250
```

To display the cell counts expressed as a proportion of the row or column totals instead of the grand total, set the margin argument to 1 for rows, 2 for columns, and 3+ for higher dimensions:

```
> prop.table(sexeyetable, margin=2)
```

```
           Male      Female
Blue   0.4000000 0.5000000
Brown  0.5000000 0.1666667
Green  0.1000000 0.3333333
```

To display percentages, multiply the whole table by 100. You can also use the round function to round all of the numbers in the table:

```
> round(prop.table(sexeyetable)*100)
```

```
       Male Female
Blue     25     19
Brown    31      6
Green     6     12
```

The addmargins function displays your table with row and column totals:

```
> addmargins(sexeyetable)
```

```
       Male Female Sum
Blue      4      3   7
Brown     5      1   6
Green     1      2   3
Sum      10      6  16
```

To add margins to just one dimension of the table, set the margin argument to 1 for rows, 2 for columns, and 3+ for higher dimensions:

```
> addmargins(sexeyetable, margin=1)
```

```
       Male Female
Blue      4      3
Brown     5      1
Green     1      2
Sum      10      6
```

Creating Tables from Count Data

So far, you have created tables by counting each row in the dataset as one observation. However, occasionally you may have a dataset in which count data has already been aggregated, such as the warpbreaks dataset (included with R).

To create a table from a data frame containing count data, use the xtabs function:

```
> xtabs(breaks~wool+tension, warpbreaks)
```

```
     tension
wool   L   M   H
   A 401 216 221
   B 254 259 169
```

If the data frame has more than one column of count data, put them inside the list function:

```
> xtabs(list(counts1, counts2)~factor1, dataset)
```

You can create a table object by assigning the output of the xtabs function to a new object name:

```
> warpbreakstable<-xtabs(breaks~wool+tension, warpbreaks)
```

To create a data frame of count data from a table object, use the as.data.frame function:

```
> as.data.frame(Titanic)
```

```
   Class    Sex   Age Survived Freq
1    1st   Male Child       No    0
2    2nd   Male Child       No    0
3    3rd   Male Child       No   35
4   Crew   Male Child       No    0
5    1st Female Child       No    0
6    2nd Female Child       No    0
7    3rd Female Child       No   17
8   Crew Female Child       No    0
9    1st   Male Adult       No  118
10   2nd   Male Adult       No  154
11   3rd   Male Adult       No  387
12  Crew   Male Adult       No  670
13   1st Female Adult       No    4
14   2nd Female Adult       No   13
15   3rd Female Adult       No   89
16  Crew Female Adult       No    3
17   1st   Male Child      Yes    5
18   2nd   Male Child      Yes   11
19   3rd   Male Child      Yes   13
20  Crew   Male Child      Yes    0
21   1st Female Child      Yes    1
22   2nd Female Child      Yes   13
23   3rd Female Child      Yes   14
24  Crew Female Child      Yes    0
25   1st   Male Adult      Yes   57
26   2nd   Male Adult      Yes   14
27   3rd   Male Adult      Yes   75
28  Crew   Male Adult      Yes  192
29   1st Female Adult      Yes  140
30   2nd Female Adult      Yes   80
31   3rd Female Adult      Yes   76
32  Crew Female Adult      Yes   20
```

Creating a Table Directly

Sometimes you may not have a dataset at all, only a table of counts. In order to be able to perform analysis such as the chi-square test (covered later in this chapter), you will need to enter your data into R as a table object.

You can create a one-dimensional table object with the as.table function. Enter the counts for each of the categories as shown here:

```
> table1D<-as.table(c(5, 21, 17, 3, 1))
```

When you view the table, you can see that R has given the categories default names of A, B, and so on.

```
> table1D
```

```
A  B  C  D  E
5 21 17  3  1
```

To overwrite the default names with the correct category names, use the row.names function:

```
> row.names(table1D)<-c("Category 1", "Category 2", "Category 3", "Category 4", "Category 5")
```

You can also enter a two-dimensional table into R in a similar way. Suppose that you want to enter the data shown in Table 6-1.

Table 6-1. *Number of subjects with and without a disease after treatment with either an active treatment or a control*

		Disease Status	
		Resolved	*Unresolved*
Group	*Active Treatment Group*	15	11
	Control Group	10	15

To create a two-dimensional table, you will first have to create a *matrix* object and then convert it to a table object. A matrix is a type of object that holds a rectangular grid of data, where all the data is of the same type (either character strings or numbers). It is a bit like a vector, except with two dimensions. To create a matrix, use the matrix function:

```
> matrix1<-matrix(c(15, 10, 11, 15), nrow=2)
```

This creates the matrix shown here:

```
> matrix1
```

```
     [,1] [,2]
[1,]   15   11
[2,]   10   15
```

As you can see, R has arranged the list of values into a two-by-two matrix. R begins by filling the first column of the matrix from top to bottom, before moving onto the second column and so on. The nrow argument tells R how many rows the matrix should have, and the number of columns required is calculated automatically. Alternatively, you can use the ncol argument to specify the number of columns.

To give proper labels to the table dimensions and levels, use the dimnames argument:

```
> matrix1<- matrix(c(15, 9, 11, 17), nrow=2, dimnames=list(Group=c("Active", "Control"),
    Disease=c("Resolved", "Unresolved")))
```

Once you have created the matrix, use the as.table function to convert it to a table object:

```
> treattable<-as.table(matrix1)
```

Alternatively, you can create the table in one step by nesting the matrix function inside the as.table function:

```
> treattable<-as.table(matrix(c(15, 9, 11, 16), nrow=2, dimnames=list(Group=c("Active", "Control"),
    Disease=c("Resolved", "Unresolved"))))
```

Chi-Square Goodness-of-Fit Test

The *chi-square goodness-of-fit* test (also known as the Pearson's chi-squared test or χ^2 test) allows you to compare categorical data with a theoretical distribution. It has the null hypothesis that the data follows the specified distribution, and the alternative hypothesis that it does not. The test is only suitable if sufficient data is available, which is commonly defined as each category having an expected frequency (under the null hypothesis) of at least five.

The test should not be confused with the *chi-square test of association* (see the next section in this chapter), which helps to determine whether two or more categorical variables are associated. The chi-square goodness-of-fit test and the chi-square test of association are both forms of the Pearson chi-square test, but they answer different questions about the data. To illustrate the difference, if you had recorded the results of a series of six-sided die rolls and wanted to use this data to determine whether your die was fair, you would use the chi-square goodness-of-fit test. The null hypothesis for the test would be that each of the six sides is equally likely to be rolled, with probability 1/6. However, if you had recorded the name of the person rolling the die as well as the result of the die roll and you wanted to determine whether there was any relationship between the result of the roll and the person that rolled the die, you would use the chi-square test of association.

■ **Note** For more details about hypothesis testing, see Chapter 10.

You can perform a chi-square goodness-of-fit test with the chisq.test function. If you have a one-dimensional table object, you can test it against the uniform distribution (i.e., against the null hypothesis that all categories are equally likely to occur), as shown here:

```
> chisq.test(tableobject)
```

If you are using raw data, nest the table function inside the chisq.test function:

```
> chisq.test(table(dataset$factor1))
```

To test the data against a different theoretical distribution, use the p argument to give a list of expected relative frequencies under the null hypothesis. You must give the same number of relative frequencies as there are levels in your table, and they must sum to one.

For example, to test the hypothesis that 10 percent of the population belong to the first category, 40 percent to the second category, 40 percent to the third category, and 10 percent to the fourth category, use the command:

```
> chisq.test(tableobject, p=c(0.1, 0.4, 0.4, 0.1))
```

EXAMPLE 6-1.
CHI-SQUARE GOODNESS-OF-FIT TEST USING THE APARTMENTS DATA

Consider the apartments dataset (available with the downloads for this book and described in Appendix C), which give details of thirty-nine one-bedroom apartments advertised for rent in a particular area of the United Kingdom in October 2012. The Price.Cat variable gives the rental price category for the apartment.

To create a table showing the number of apartments in each price category, use the command:

```
> table(apartments$Price.Cat)
```

£500-550	£551-600	£601-650	£651+
5	13	8	6

It is believed that 20 percent of the one-bedroom apartments in this area have a rental price less than £550, 30 percent have a price between £551 and £600, 30 percent have a price between £601 and £650, and 20 percent have a rental price greater than £650.

Suppose that you want to use a chi-square goodness-of-fit test to determine whether the data is consistent with the hypothesized price distribution. The test has the null hypothesis that the described price distribution is correct, and the alternative hypothesis that it is not. A significance level of 0.05 is used.

To perform the test, use the command:

```
> chisq.test(table(apartments$Price.Cat), p=c(0.2, 0.3, 0.3, 0.2))
```

This gives the output:

```
        Chi-squared test for given probabilities

data:  table(apartments$Price.Cat)
X-squared = 1.8021, df = 3, p-value = 0.6145
```

The p-value of 0.6145 is not less than the significance level of 0.05, so we cannot reject the null hypothesis. This means that the data is consistent with the hypothesised price distribution.

Tests of Association Between Categorical Variables

In the same way that we can look at the association between continuous variables using statistics such as covariance and correlation, there are methods available to help you to determine whether there is an association between categorical variables. Two of these are the chi-square test of association and Fisher's exact test. These tests answer the same question but are suitable in different situations.

The chi-square test of association is used to test for association between two or more variables, and can be used regardless of how many levels there are in each variable. However, the test is only suitable when there is plenty of data available.

By contrast, Fisher's exact test can only be used to look for association between two categorical variables which each have two levels. However, it is suitable even when very little data is available.

The following sections explain how to perform these tests in R.

Chi-Square Test of Association

The *chi-square test of association* (sometimes called the chi-square test of independence) helps to determine whether two or more categorical variables are associated. The test has the null hypothesis that the variables are independent and the alternative hypothesis that they are not independent (i.e., at least two of the variables are associated) The test is only suitable if there is sufficient data, which is commonly defined as all table cells having expected counts (under the null hypothesis) of at least five.

The summary function performs a chi-square test of association when given a table object as input. If you have already created a table object, use the summary function directly:

```
> summary(tableobject)
```

If you have raw data, nest the table function inside the summary function:

```
> summary(table(dataset$var1, dataset$var2, dataset$var3))
```

EXAMPLE 6-2.
CHI-SQUARE TEST OF ASSOCIATION USING PEOPLE2 DATA

Suppose that you want to use a chi-square test of association to determine whether sex and eye colour are associated, using the people2 dataset.

If you still have the sexeyetable object created earlier in this chapter in the "Frequency Tables" section, use the command:

```
> summary(sexeyetable)
```

Alternatively, you can perform the test using the raw data:

```
> summary(table(people2$Sex, people2$Eye.Colour))
```

```
Number of cases in table: 16
Number of factors: 2
Test for independence of all factors:
        Chisq = 2.2857, df = 2, p-value = 0.3189
        Chi-squared approximation may be incorrect
```

As the p-value of 0.3189 is not less than the significance level of 0.05, we cannot reject the null hypothesis that sex and eye color are independent. This means that there is no evidence of an association between the two.

The warning `Chi-squared approximation may be incorrect` tells us that as some cells have expected counts less than five. This means that the results may be unreliable and should be interpreted with caution. A discussion of chi-square tests with small expected counts can be found on page 39 of *The Analysis of Contingency Tables*, Second Edition, by B. S. Everitt (Chapman & Hall/CRC, 1992).

EXAMPLE 6-3.
CHI-SQUARE TEST OF ASSOCIATION USING THE ESOPH DATA

Consider the `esoph` dataset, which is included with R. The dataset gives the results of a case-control study of oesophageal cancer. You can view more details by entering `help(esoph)`. The `agegp`, `alcgp` and `tobgp` variables list categories for the subjects' age, alcohol consumption, and smoking habits. The variables `ncases` and `ncontrols` give the number of subjects with and without oesophageal cancer that fall into each of these categories.

Suppose that you want to use a chi-square test of association to determine where there is any association between smoking habits and oesophageal cancer. The test has the null hypothesis that smoking habits and oesophageal cancer are independent, and the alternative hypothesis that they are associated. A significance level of 0.05 will be used.

As the dataset contains data that is already expressed as counts, you can use the `xtabs` function to create a table giving the number of subjects with and without oesophageal cancer that fall into each category of smoking habits:

```
> tobacco<-xtabs(cbind(ncases, ncontrols)~tobgp, esoph)
> tobacco
```

tobgp	ncases	ncontrols
0-9g/day	78	525
10-19	58	236
20-29	33	132
30+	31	82

To perform the test, use the command:

```
> summary(tobacco)
```

```
Call: xtabs(formula = cbind(ncases, ncontrols) ~ tobgp, data = esoph)
Number of cases in table: 1175
Number of factors: 2
Test for independence of all factors:
        Chisq = 18.363, df = 3, p-value = 0.0003702
```

As the p-value of 0.0003702 is less than the significance level of 0.05, we can reject the null hypothesis that smoking category and oesophageal cancer are independent, in favour of the alternative hypothesis that the two are associated. This means that there is evidence of a relationship between smoking habits and oesophageal cancer.

Fisher's Exact Test

The Fisher's exact test is used to test for association between two categorical variables that each have two levels. Unlike the chi-square test of association, it can be used even when very little data is available. The test has the null hypothesis that the two variables are independent and the alternative hypothesis that they are not independent.

You can perform a Fisher's exact test with the fisher.test function. You can use the function with a two-by-two table object:

```
> fisher.test(tableobject)
```

You can also use raw data:

```
> fisher.test(dataset$var1, dataset$var2)
```

The test results are accompanied by a 95 percent confidence interval for the odds ratio. You can change the size of the interval with the conf.level argument:

```
> fisher.test(dataset$var1, dataset$var2, conf.level=0.99)
```

EXAMPLE 6-4.
FISHER'S EXACT TEST USING PEOPLE2 DATA

Using the table function, we can see there are more left-handed women than left-handed men in the people2 dataset:

```
> table(people2$Sex, people2$Handedness)
```

```
       Left Right
Male      0    9
Female    2    2
```

Suppose you want to perform a Fisher's exact test to determine whether there is any statistically significant relationship between sex and handedness. The test has the null hypothesis that sex and handedness are independent, and the alternative hypothesis that they are associated. A significance level of 0.05 is used.

To perform the test, use the command:

```
> fisher.test(people2$Sex, people2$Handedness)
```

This gives the output:

```
        Fisher's Exact Test for Count Data

data:  people2$Sex and people2$Handedness
p-value = 0.07692
alternative hypothesis: true odds ratio is not equal to 1
```

```
95 percent confidence interval:
 0.000000 2.098087
sample estimates:
odds ratio
         0
```

Because the p-value of 0.07692 is not less than the significance level of 0.05, we cannot reject the null hypothesis that sex and handedness are independent. This means that there is no evidence of a relationship between sex and handedness.

Proportions Test

A test of proportions allows you to compare the proportion of observations with a given outcome or attribute (referred to as a *success*) across two or more groups of observations to determine whether they are significantly different. The null hypothesis for the test is that the probability of a success is the same for all of the groups, and the alternative hypothesis is that the probability of success is not the same for all of the groups.

The test can be applied to an n-by-two contingency table, where the two columns give the number of successes and failures, and the n rows give the number that fall into each group.

You can perform a test of proportions with the prop.test function. If your data is already saved in a table object, you can use the command:

```
> prop.test(tableobject)
```

For the command to work, your table must have exactly two columns, and the rows of the table should show the different groups that you want to compare. If your table is the wrong way around, so that the columns give groups and the rows give successes and failures, you can use the t function to transpose the table:

```
> prop.test(t(tableobject))
```

If you don't have a table object, you can perform the test using raw data by nesting the table function inside the prop.test function:

```
>prop.test(table(dataset$groups, dataset$outcome))
```

Alternatively, you can perform the test using raw count data:

```
> prop.test(c(success1, success2), c(n1, n2))
```

where success1 and success2 are the number of successes in Group 1 and Group 2, respectively, and n1 and n2 are the total number of observations in Group 1 and Group 2, respectively.

EXAMPLE 6-5.
PROPORTIONS TEST

Consider the data shown in Table 6-1 (under "Creating Tables Directly"), which gives the number of patients with resolved and unresolved disease after treatment with either an active treatment or a control. If you still have the treattable table object, you can view the contents:

```
> treattable
```

```
        Disease
Group     Resolved Unresolved
  Active        15         11
  Control        8         17
```

To see the proportion with resolved and unresolved disease for each treatment group, use the command:

```
> prop.table(treattable, margin=1)
```

```
        Disease
Group      Resolved Unresolved
  Active  0.5769231  0.4230769
  Control 0.3200000  0.6800000
```

From the output, you can see that the proportion of subjects with resolved disease is 58 percent in the treatment group and 32 percent in the control group. Suppose that you want to use a test of proportions to help determine whether this difference is significant or whether it is just the result of random variation. A significance level of 0.05 will be used.

To perform the test, enter the command:

```
> prop.test(treattable)
```

Alternatively, you can perform the test using the raw data:

```
> prop.test(c(15,8), c(26,25))
```

This produces the following output:

```
        2-sample test for equality of proportions with continuity correction

data:  treattable
X-squared = 1.6153, df = 1, p-value = 0.2038
alternative hypothesis: two.sided
95 percent confidence interval:
 -0.08963846  0.52348461
sample estimates:
   prop 1    prop 2
0.5769231 0.3600000
```

From the output, you can see that the p-value of 0.2038 is not less than the significance level of 0.05, so we cannot reject the null hypothesis that the proportion of patients with resolved disease is the same in both treatment groups. This means that there is no evidence of a difference in the probability of disease resolution for the active treatment and control.

Summary

You should be able to create frequency and contingency tables to summarize categorical data and be able to present them with marginal sums or as percentages if required. You should be able to compare your categorical data to a hypothesised distribution using the chi-square goodness-of-fit test. You should also be able to use the chi-square test of association or Fisher's exact test to look for association between two categorical variables. Finally, you should be able to use a test to compare two or more proportions.

This table summarizes the most important commands covered.

Task	Command
Contingency table	`table(dataset$factor1, dataset$factor2)`
Compact table	`ftable(tableobject)`
Proportions table	`prop.table(tableobject)`
Table with margins	`addmargins(tableobject)`
Chi-square goodness-of-fit test	`chisq.test(tableobject, p=c(p1, p2, pN))` `chisq.test(table(dataset$factor1), p=c(p1, p2, pN))`
Chi-square test of association	`summary(tableobject)` `summary(table(dataset$factor1, dataset$factor2))`
Fisher's exact test	`fisher.test(tableobject)` `fisher.test(dataset$factor1, dataset$factor2)`
Proportions test	`prop.test(tableobject)` `prop.test(table(dataset$groups, dataset$outcome))`

Now that you have learned how to summarize continuous and categorical variables, we can move on to the next chapter, in which you will learn about probability distributions.

■ ■ ■

CHAPTER 7

Probability Distributions

This chapter looks at the topic of probability distributions such as the normal distribution, exponential distribution, binomial distribution and Poisson distribution. There are several ways of describing a probability distribution, which include giving its probability density function (or probability mass function for discrete distributions), its cumulative distribution function or its inverse cumulative distribution function. These functions are useful in the same way as a set of statistical tables, as they allow you to calculate probabilities and quantiles for a given distribution.

Random number generation is also covered in this chapter. Random numbers allow you to simulate data, which is useful for trying out statistical techniques in the absence of organic data.

You will learn how to:

- calculate the probability that a random variable will take a given value or fall within a given range of values

- find the range of values within which a given percentage of the population falls

- generate random numbers

Probability Distributions in R

R has functions for all of the well-known probability distributions. For each distribution, there are the following four functions:

Probability density function or probability mass function (prefix d) For discrete distributions, you can use the probability mass function (pmf) to answer questions of the type *"What is the probability that the outcome will be equal to x?"*

Cumulative distribution function (prefix p) You can use the cumulative density function (cdf) to answer questions of the type *"If we were to randomly select a member of a given population, what is the probability that it will have a value less than x, or a value between x and y?"*

Inverse cumulative distribution function (prefix q) Use the inverse cdf to answer questions such as *"Which value do x% of the population fall below,"* or *"What range of values do x% of the population fall within?"*

Random number generator (prefix r) Use the random number generator to simulate a random sample from a given distribution.

The function names are formed from a prefix (either d, p, q, or r) and a suffix for the relevant distribution, taken from Table 7-1. For example, the cumulative distribution function for the normal distribution is called pnorm, and the random number generator for the exponential distribution is called rexp.

Table 7-1. *Standard probability distributions included with R. Combine the distribution suffix with prefix: d for the pdf; p for the cdf; q for the inverse cdf; and r for the random number generator*

Distribution	Function suffix
Beta	beta
Binomial	binom
Cauchy	cauchy
Chi-square	chisq
Exponential	exp
F	f
Gamma	gamma
Geometric	geom
Hypergeometric	hyper
Logistic	logis
Log normal	lnorm
Multinomial	multinom
Negative binomial	nbinom
Normal	norm
Poisson	pois
Student's t distribution	t
Uniform	unif
Weibull	weibull

THE PROBABILITY DENSITY FUNCTION AND CUMULATIVE DISTRIBUTION FUNCTION

The probability density function (pdf) and cumulative distribution function (cdf) are two ways of specifying the probability distribution of a random variable.

The pdf is denoted f(x) and gives the relative likelihood that the value of the random variable will be equal to x. The total area under the curve is equal to 1.

The cdf is denoted F(x) and gives the probability that the value of a random variable will be less than or equal to x. The value of the cdf at x is equal to the area under the curve of the pdf between -∞ and x. The cdf takes values between 0 and 1.

Figure 7-1 shows the pdf and cdf for the standard normal distribution.

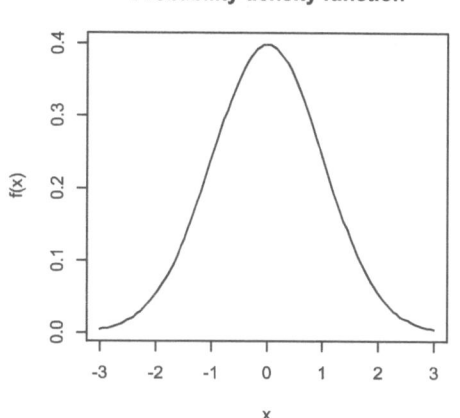
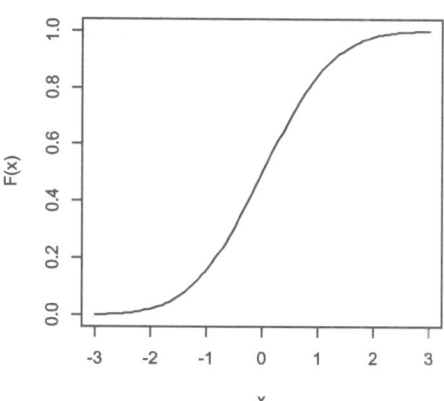

Figure 7-1. *The probability density function and cumulative distribution function for the standard normal distribution*

Probability Density Functions and Probability Mass Functions

For discrete distributions, use the probability mass function (pmf) to find the probability that an outcome or observation is equal to a given value x. The name of the function is formed by combining the prefix d with the relevant suffix from Table 7-1.

When using the pmf, the first argument that the function requires is always the possible outcome of interest (x). This is followed by any parameters relevant to the distribution. You can check these with the help function. The examples in this section demonstrate how to use the pmfs of the binomial and Poisson distributions.

For continuous distributions, the probability density function (pdf) gives the relative likelihood of x, which has no direct interpretation. However, you can still find the value of the pdf in the same way as for discrete distributions.

In Chapter 8, in the "Plotting a Function" section, you will learn how you can use the curve function to plot the pdf of a probability distribution.

EXAMPLE 7-1.
PROBABILITY MASS FUNCTION FOR THE BINOMIAL DISTRIBUTION

Suppose that a fair die is rolled 10 times. What is the probability of throwing exactly two sixes?

You can answer the question using the dbinom function:

```
> dbinom(2, 10, 1/6)
```

```
[1] 0.29071
```

The probability of throwing two sixes is approximately 0.29 or 29 percent.

EXAMPLE 7-2.
PROBABILITY MASS FUNCTION FOR THE POISSON DISTRIBUTION

The number of lobster ordered in a restaurant on a given day is known to follow a Poisson distribution with a mean of 20. What is the probability that exactly eighteen lobsters will be ordered tomorrow?

You can answer the question with the dpois function:

```
> dpois(18, 20)
```

[1] 0.08439355

The probability that exactly eighteen lobsters are ordered is 8.4 percent.

EXAMPLE 7-3.
PROBABILITY DENSITY FUNCTION FOR THE NORMAL DISTRIBUTION

To find the value of the pdf at x=2.5 for a normal distribution with a mean of 5 and a standard deviation of 2, use the command:

```
> dnorm(2.5, mean=5, sd=2)
```

[1] 0.09132454

The value of the pdf at x = 2.5 is 0.091.

Finding Probabilities

To find the probability that a randomly selected member of a population will have a value less than or equal to a given value x, use the cumulative distribution function (cdf) for the appropriate distribution. The name of the function is formed by combining the prefix p with the relevant suffix from Table 7-1.

For the normal distribution, use the pnorm function. For a standard normal distribution (with a mean of 0 and a standard deviation of 1) the function does not require any additional arguments. For example, to find the probability that a randomly selected value will be less than or equal to 2.5, use the command:

```
> pnorm(2.5)
```

[1] 0.9937903

From the output, you can see that the probability is 0.9937903, or approximately 99 percent.

To find a probability for a nonstandard normal distribution, add the mean and sd arguments. For example, if a random variable is known to be normally distributed with a mean of 5 and a standard deviation of 2 and you wish to find the probability that a randomly selected member will be no more than 6, use the command:

```
> pnorm(6, mean=5, sd=2)
[1] 0.6914625
```

To find the complementary probability that the value will be *greater* than 6, set the lower.tail argument to F:

```
> pnorm(6, 5, 2, lower.tail=F)
```

```
[1] 0.3085375
```

This is equivalent to the command:

```
> 1-pnorm(6, 5, 2)
```

```
[1] 0.3085375
```

You can find probabilities for other distributions by substituting norm for the relevant distribution suffix from Table 7-1. You will also need to change the mean and sd arguments for the parameters relevant to the distribution. Use the help function to check the parameters for the function that you are using.

EXAMPLE 7-4.
FINDING PROBABILITIES FOR THE NORMAL DISTRIBUTION

Suppose the height of men in the United Kingdom is known to be normally distributed with a mean of 177 centimeters and a standard deviation of 10 centimeters. If you were to select a man from the United Kingdom population at random, what is the probability that he would be more than 200 centimeters tall?

To answer the question, use the command:

```
> pnorm(200, 177, 10, lower.tail=F)
```

```
[1] 0.01072411
```

From the output, you can see that the probability is approximately 0.011, or 1.1 percent.

What is the probability that he would be less than 150 centimeters tall?

```
> pnorm(150, 177, 10)
```

```
[1] 0.003466974
```

The probability is 0.0035, or 0.35 percent.

EXAMPLE 7-5.
FINDING PROBABILITIES FOR THE BINOMIAL DISTRIBUTION

If you were to roll a fair six-sided die 100 times, what is the probability of rolling a six no more than 10 times?

The number of sixes in 100 dice rolls follows a binomial distribution, so you can answer the question with the pbinom function

```
> pbinom(10, 100, 1/6)
```

[1] 0.04269568

From the output, you can see that the probability of rolling no more than 10 sixes is 0.043 (4.3%).

What is the probability of rolling a six more than 20 times?

```
> pbinom(20, 100, 1/6, lower.tail=F)
```

[1] 0.1518878

The probability of rolling more than 20 sixes is approximately 0.15, or 15 percent.

EXAMPLE 7-6.
FINDING PROBABILITIES FOR THE EXPONENTIAL DISTRIBUTION

Malfunctions in a particular type of electronic device are known to follow an exponential distribution with a mean time of 24 months until the device malfunctions. What is the probability that a randomly selected device will malfunction within the first 6 months?

You can answer the question using the pexp function

```
> pexp(6, 1/24)
```

[1] 0.2211992

The probability of malfunction within six months is 0.22 (22%).

What is the probability that a randomly selected device will last more than 5 years (60 months) without malfunction?

```
> pexp(60, 1/24, lower.tail=F)
```

[1] 0.082085

The probability that it will last more than 5 years is approximately 0.08, or 8 percent.

Finding Quantiles

To find the value that a given percentage of a population falls above or below, or the range of values within which a given percentage of a population lies, use the inverse cdf for the appropriate distribution. The name of the function is formed by combining the prefix q with the relevant suffix from Table 7-1.

To find quantiles for the normal distribution, use the qnorm function. For a standard normal distribution, use the function without any additional arguments. For example, to find the value below which 95 percent of values fall, use the command:

```
> qnorm(0.95)
```

```
[1] 1.644854
```

For nonstandard normal distributions, use the mean and sd arguments to specify the parameters for the distribution. For example, suppose that a variable is known to be normally distributed with a mean of 5 and standard deviation of 2. To find the value below which 95 percent of the population falls, use the command:

```
> qnorm(0.95, mean=5, sd=2)
```

```
[1] 8.289707
```

To find the value *above* which 95 percent of the population falls, set the lower.tail argument to F:

```
> qnorm(0.95, 5, 2, lower.tail=F)
```

```
[1] 1.710293
```

For other standard probability distributions, you must replace the mean and sd arguments with other parameters relevant to the distribution that you are using. Use the help function to see what these are.

EXAMPLE 7-7.
FINDING QUANTILES FOR THE NORMAL DISTRIBUTION

A manufacturer of a special type of one-size glove wants to design the glove to fit at least 99 percent of the population. Hand span is known to be normally distributed with a mean of 195 millimeters and a standard deviation of 17 millimeters. What range of hand spans must the glove accommodate?

To find the value below which 0.5 percent of the population falls, use the command:

```
> qnorm(0.005, 195, 17)
```

```
[1] 151.2109
```

Similarly, to find the value above which 0.5 percent of the population falls, use the command:

```
> qnorm(0.005, 195, 17, lower.tail=F)
```

```
[1] 238.7891
```

The remaining 99 percent of the population falls between these two values. So, to accommodate 99 percent of the population, the gloves must be designed to fit hands with a span between 151 and 239 millimeters.

EXAMPLE 7-8.
FINDING QUANTILES FOR THE EXPONENTIAL DISTRIBUTION

Malfunctions in a particular type of electronic device are known to follow an exponential distribution with a mean time of 24 months until the device malfunctions. After how many months will 40 percent of the devices already have malfunctioned?

To find the length of time within which 40 percent of devices will have malfunctioned, use the command:

```
> qexp(0.4, 1/24)
```

```
[1] 12.25981
```

So 40 percent of devices will malfunction within 12.3 months.

EXAMPLE 7-9.
FINDING QUANTILES FOR THE POISSON DISTRIBUTION

The number of lobsters ordered in a restaurant on a given day is known to follow a Poisson distribution with a mean of 20. If the manager wants to be able to satisfy all requests for lobster on at least 80 percent of days, how many lobster should they order each day?

To find the number of lobster requests that will not be exceeded on 80 percent of days, use the command:

```
> qpois(0.8, 20)
```

```
[1] 24
```

By ordering 24 lobsters per day, the restaurant will be able to satisfy all requests for lobster on at least 80 percent of days.

Generating Random Numbers

R has a set of functions that allow you to generate random numbers from any of the well-known probability distributions. This is useful for simulating data.

The functions' names are formed from the prefix r and the relevant suffix from Table 7-1. The first argument required by the functions is the quantity of random numbers to generate. This is followed by any parameters relevant to the distribution, which you can check using the help function.

For example, this command generates 100 random numbers from a standard normal distribution and saves them in an object named vector1:

```
> vector1<-rnorm(100)
```

To generate random numbers from a nonstandard normal distribution, add the mean and sd arguments. For example, to generate 100 random numbers from a normal distribution with a mean of 27.3 and a standard deviation of 4.1, use the command:

```
> vector2<-rnorm(100, 27.3, 4.1)
```

To generate a simple random sample from a range of numbers, you can use the sample function. For example, to select 20 random numbers between 1 and 100 without replacement, use the command:

```
> sample(1:100, 20)
```

```
[1] 58 76 87 14 95 68 98  2 47 96 12 49 44 21 23 34 84  6 29  5
```

To sample with replacement, set the replace argument to T:

```
> sample(1:100, 20, replace=T)
```

EXAMPLE 7-10.
GENERATING RANDOM NUMBERS FROM A NORMAL DISTRIBUTION

Hand span in a particular population is known to be normally distributed with a mean of 195 millimeters and a standard deviation of 17 millimeters. To simulate the hand spans of three randomly selected people, use the command:

```
> rnorm(3, 195, 17)
```

```
[1] 186.376 172.164 195.504
```

EXAMPLE 7-11.
GENERATING RANDOM NUMBERS FROM A BINOMIAL DISTRIBUTION

To simulate the number of sixes thrown in 10 rolls of a fair die, use the command:

```
> rbinom(1, 10, 1/6)
```

[1] 3

EXAMPLE 7-12.
GENERATING RANDOM NUMBERS FROM A POISSON DISTRIBUTION

The number of lobsters ordered on any given day in a restaurant follows a Poisson distribution with a mean of 20. To simulate the number of lobsters ordered over a seven-day period, use the command:

```
> rpois(7, 20)
```

[1] 19 10 13 23 21 13 25

EXAMPLE 7-13.
GENERATING RANDOM NUMBERS FROM AN EXPONENTIAL DISTRIBUTION

Malfunctions in a particular type of electronic device are known to follow an exponential distribution with a mean time of 24 months until the device malfunctions.

To simulate the time to malfunction for ten randomly selected devices, use the command:

```
> rexp(10, 1/24)
```

```
[1]  7.7949626  1.4280596 63.6650676 33.3343910  0.5911718 46.2797640
[7] 16.4355239 38.1404491 12.2637182 10.9453535
```

Summary

You should now be able to use the appropriate probability density function or probability mass function, cumulative distribution function, and inverse cumulative distribution function to calculate probabilities and quantiles for all of the commonly used distributions. You should also be able to generate a random sample from a given distribution.

This table shows the most important commands covered, using the normal distribution to illustrate.

Task	Command
Find the value of the density function at value x	dnorm(*x*, *mean*, *sd*)
Obtain P(X ≤ x)	pnorm(*x*, *mean*, *sd*)
Obtain x where P(X ≤ x)=p	qnorm(*p*, *mean*, *sd*)
Generate n random numbers from a normal distribution	rnorm(*n*, *mean*, *sd*)

In the next chapter, you will learn how to visualize your data using plots.

CHAPTER 8

■ ■ ■

Creating Plots

It is always a good idea to include a plot as part of your statistical analysis. Plots allow for a more intuitive grasp of the data, making them ideal for presenting your results to those without statistical expertise. They also make it easier to spot features of the data such as outliers or a bimodal distribution, which you may overlook when using other methods.

One of the strong points of R is that it makes it easy to produce excellent quality, fully customizable plots and statistical graphics. This chapter explains how to create the most popular types, which are:

- simple line plots
- histograms
- normal probability plots
- stem-and-leaf plots
- bar charts
- pie charts
- scatter plots
- scatterplot matrices
- box plots
- function plots

This chapter concentrates on creating plots using the default settings. In Chapter 9 you will learn how to make your plots more presentable by changing titles, labels, colors, and other aspects of the plot's appearance.

This chapter uses the trees and iris datasets (which are included with R), the people2 dataset (available from the website), and the sexeyetable table object (created in Chapter 6).

Simple Plots

To create a basic plot of a continuous variable against the observation number, use the plot function. For example, to plot the Height variable from the trees dataset, use the command:

```
> plot(trees$Height)
```

When you create a plot, it is displayed in a new window called the graphics device (or for Mac users, the Quartz device). Figure 8-1 shows how the plot looks.

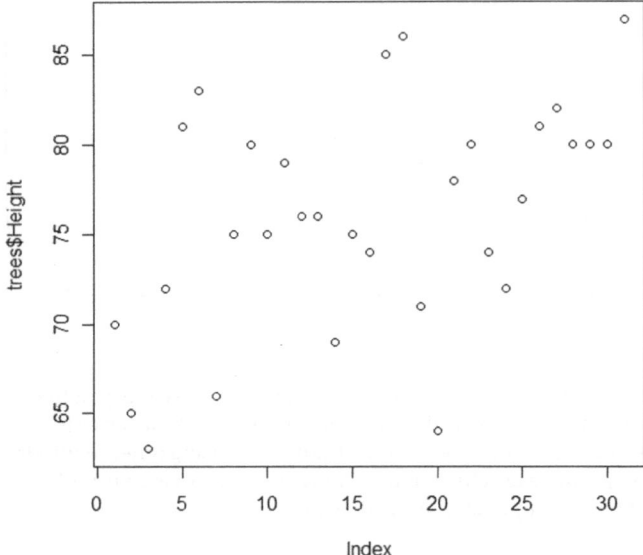

Figure 8-1. *Simple plot created with the* plot *function*

The plot function does not just create basic one-dimensional plots. The type of plot created depends on the type and number of variables you give as input. You will see it used in different parts of this book to create other types of plots, including bar charts and scatter plots.

By default, R uses symbols to plot the data values. To use lines instead of symbols, set the type argument to "l":

```
> plot(trees$Height, type="l")
```

Other possible values for the type argument include "b" for both lines and symbols, "h" for vertical lines, and "s" for steps. Figure 8-2 shows how the plot looks when these values are used. Depending on the nature of your data, some of these options will be more suitable than others.

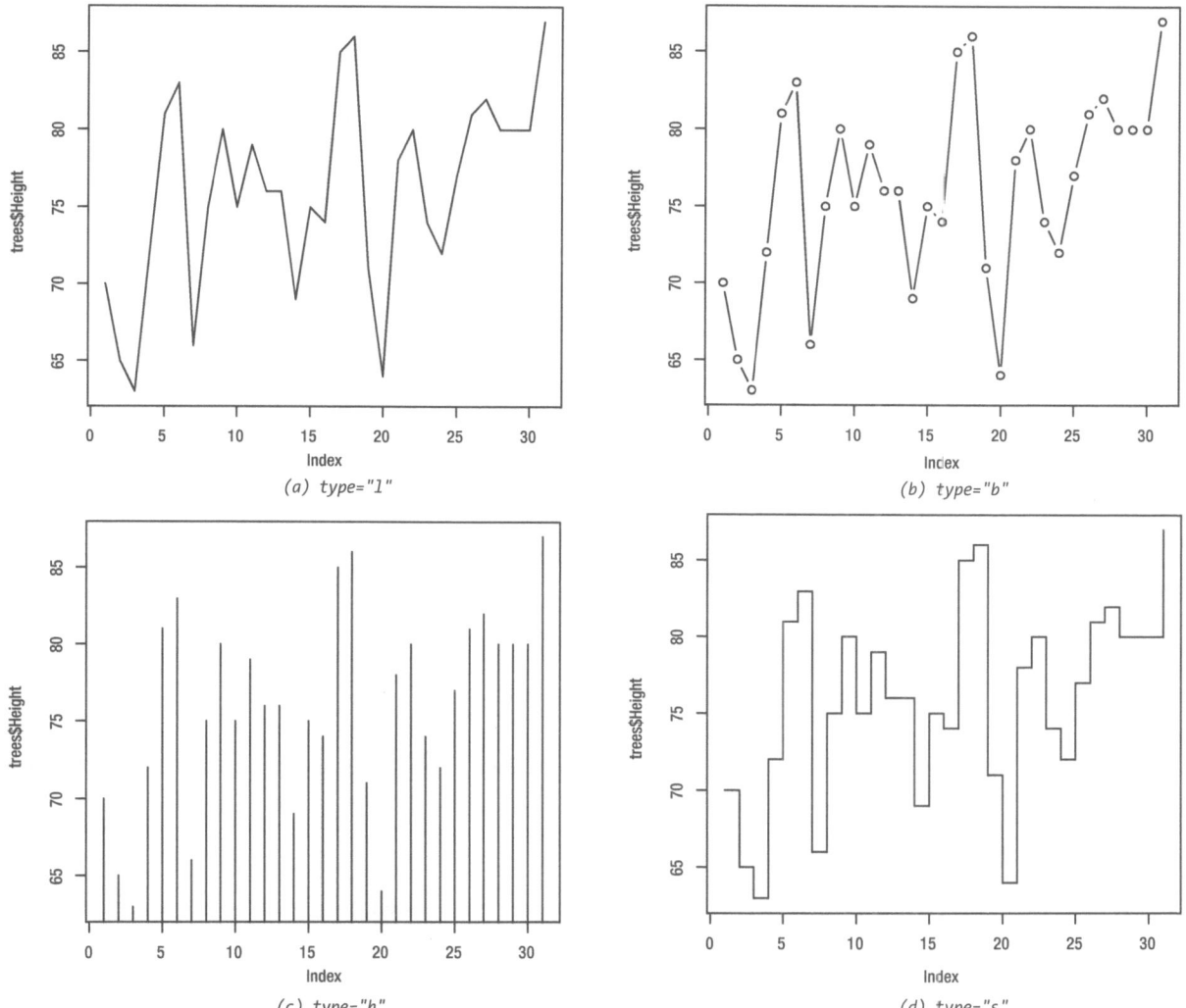

Figure 8-2. *Simple plot showing some of the options for the* type *argument*

Histograms

A histogram is a plot for a continuous variable that allows you to assess its probability distribution. To create a histogram, the range of the data is divided into a number of intervals and the number of observations that fall into each interval is counted. A histogram can either show the frequency for each interval directly, or it may show the density (i.e., the frequency is scaled so that the total area of the histogram is equal to one).

You can create a histogram with the hist function, as shown below for the Height variable. Figure 8-3 shows the result.

```
> hist(trees$Height)
```

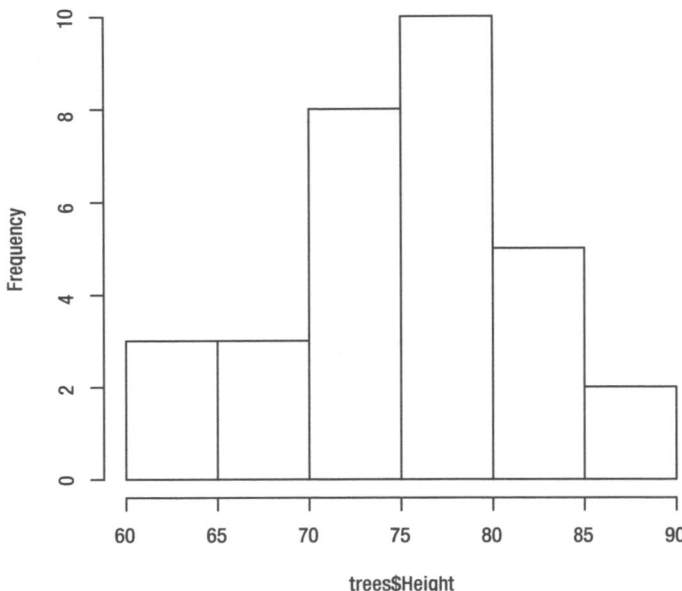

Histogram of trees$Height

Figure 8-3. Histograms of the Height variable from the trees dataset

R automatically selects a suitable number of bars for the histogram. If you prefer, you can specify the number of bars with the breaks argument:

```
> hist(dataset$variable, breaks=15)
```

By default, R creates a histogram of frequencies. To create a histogram of densities (so that the total area of the histogram is equal to one), set the freq argument to F:

```
> hist(dataset$variable, freq=F)
```

You can use the curve function to fit a normal distribution curve to the data. This allows you to see how well the data fits the normal distribution. Use the curve function directly after the hist function, while the histogram is still displayed in the graphics device. Adding a density curve is only appropriate for a histogram of densities, so remember to set the freq argument to F:

```
> hist(trees$Height, freq=F)
> curve(dnorm(x, mean(trees$Height), sd(trees$Height)), add=T)
```

The result is shown in Figure 8-4.

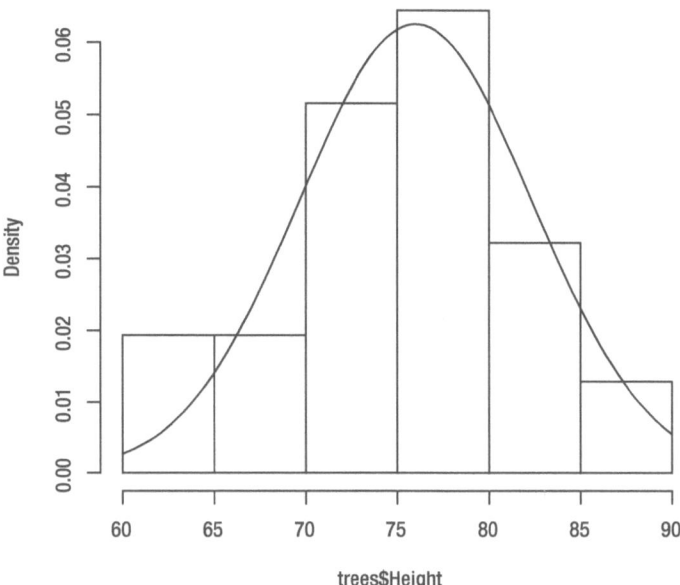

Histogram of trees$Height

Figure 8-4. *Histogram of the* Height *variable, with normal distribution curve superimposed*

If the variable has any missing data values, remember to set the na.rm argument to T for the mean and sd functions:

```
> hist(dataset$variable, freq=F)
> curve(dnorm(x, mean(dataset$variable, na.rm=T), sd(dataset$variable, na.rm=T)), add=T)
```

The curve function is discussed in more detail later in this chapter in the "Plotting a Function" section, and in Chapter 9 under "Adding a Mathematical Function Curve." The dnorm function is covered in Chapter 7 under "Probability Density Functions and Probability Mass Functions."

Normal Probability Plots

A normal probability plot is a plot for a continuous variable that helps to determine whether a sample is drawn from a normal distribution. If the data is drawn from a normal distribution, the points will fall approximately in a straight line. If the data points deviate from a straight line in any systematic way, it suggests that the data is not drawn from a normal distribution.

You can create a normal probability plot using the qqnorm function, as shown for the Height variable:

```
> qqnorm(trees$Height)
```

You can also add a reference line to the plot, which makes it easier to determine whether the data points are falling into a straight line. To add a reference line, use the qqline function directly after the qqnorm function:

```
> qqnorm(trees$Height)
> qqline(trees$Height)
```

The result is shown in Figure 8-5.

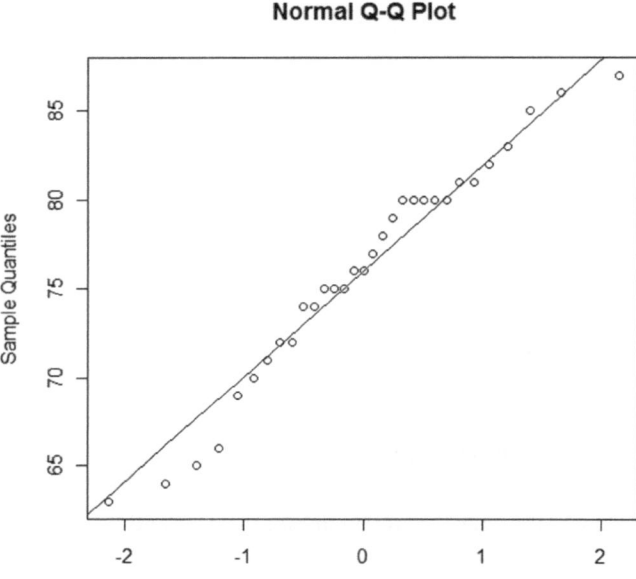

Figure 8-5. *Normal probability plot of the Height variable*

There is also a function called qqplot, which allows you to create quantile plots for comparing data with other standard probability distributions in addition to the normal distribution. Enter help(qqplot) for more details.

INTERPRETING THE NORMAL PROBABILITY PLOT

The way in which the data points fall around the straight line tells you something about the shape of the distribution relative to the normal distribution. Table 8-1 shows some patterns you might see and how to interpret them.

Table 8-1. *Patterns Seen in the Normal Probability Plot and Their Interpretations*

Normal probability plot	Corresponding histogram	Pattern and interpretation
		Pattern: Data points curve from above the line to below the line and then back to above the line. **Interpretation:** Data has a positive skew (is right-skewed).
		Pattern: Data points curve from below the line to above the line and then back to below the line. **Interpretation:** Data has a negative skew (is left-skewed).
		Pattern: Data points fall below the line toward the left and above the line toward the right. **Interpretation:** Data has a sharper peek and fatter tails relative to the normal distribution (positive excess kurtosis).

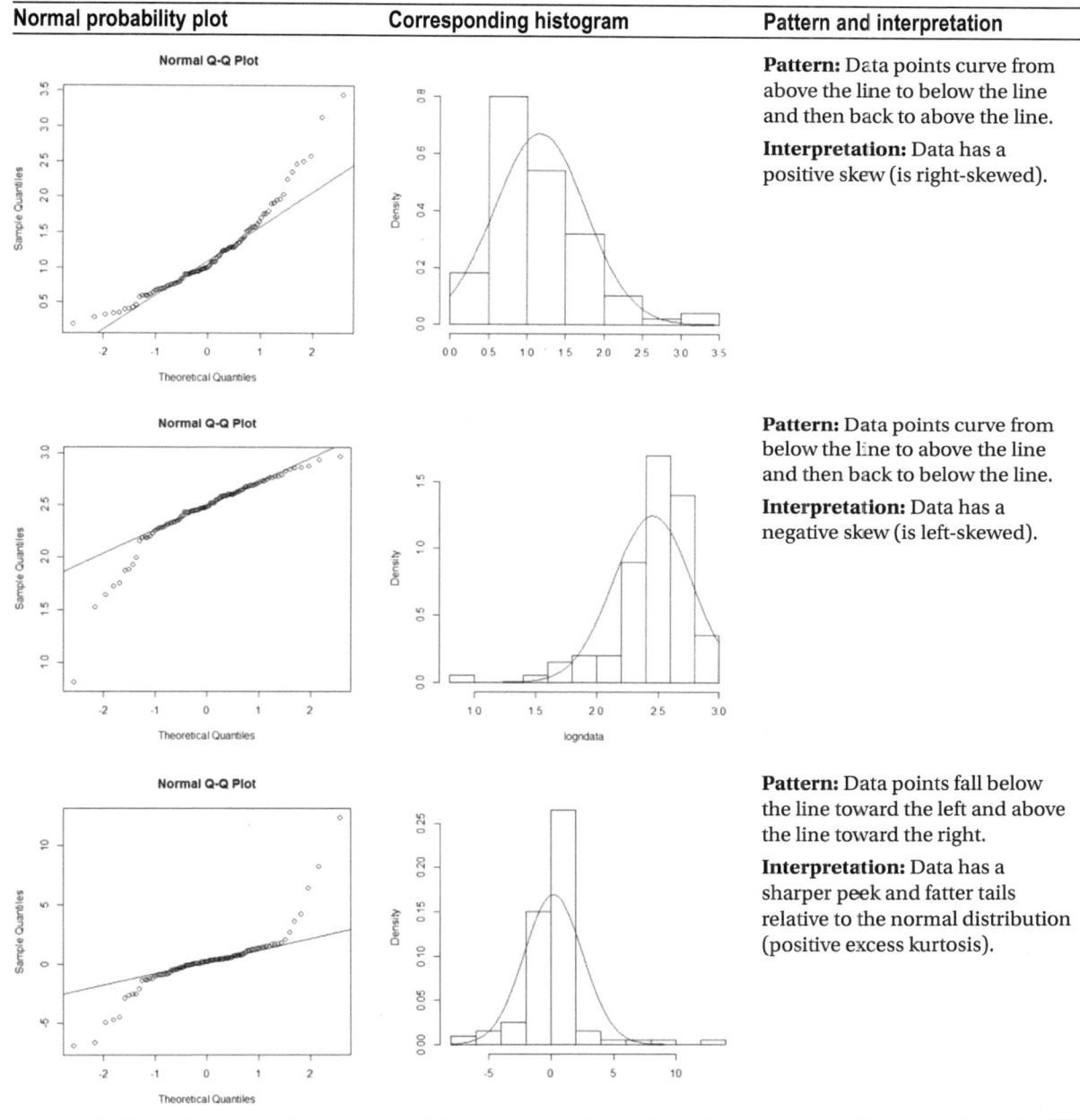

(continued)

Table 8-1. (*continued*)

Normal probability plot	Corresponding histogram	Pattern and interpretation
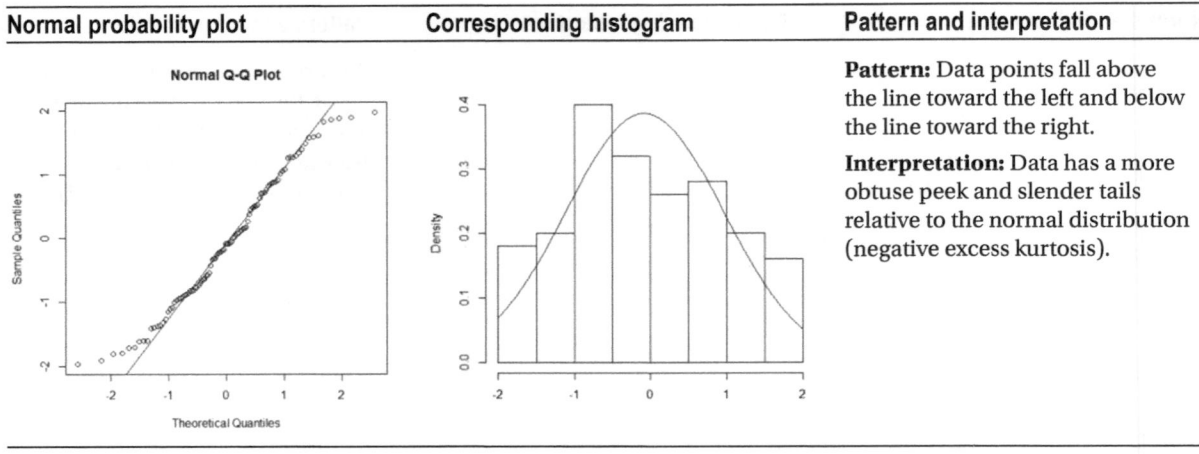		**Pattern:** Data points fall above the line toward the left and below the line toward the right. **Interpretation:** Data has a more obtuse peek and slender tails relative to the normal distribution (negative excess kurtosis).

Stem-and-Leaf Plots

The stem-and-leaf plot (or stemplot) is another popular plot for a continuous variable. It is similar to a histogram, but all of the data values can be read from the plot.

You can create a stem-and-leaf plot with the stem function, as shown here for the Volume variable:

```
> stem(trees$Volume)
```

In R, the stem-and-leaf plot is actually a semigraphical technique rather than a true plot. The output is displayed in the command window rather than the graphics device.

```
The decimal point is 1 digit(s) to the right of the |

1 | 00066899
2 | 00111234567
3 | 24568
4 | 3
5 | 12568
6 |
7 | 7
```

Bar Charts

A bar chart is a plot for summarizing categorical data. A simple bar chart summarizes one categorical variable by displaying the number of observations in each category. Grouped and stacked bar charts summarize two categorical variables by displaying the number of observations for each combination of categories.

You can create bar charts with the plot and barplot functions. If you have raw data such as the Eye.Color variable in the people2 dataset, use the plot function as shown below. The result is given in Figure 8-6.

```
> plot(people2$Eye.Color)
```

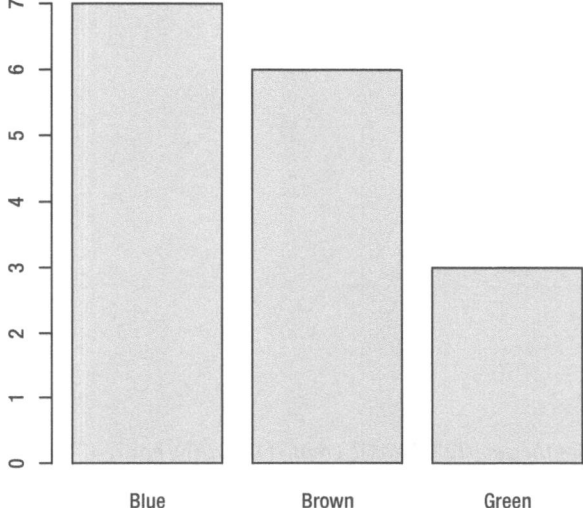

Figure 8-6. *Bar chart of the* Eye.Color *variable from the* people2 *dataset*

When creating a bar chart from raw data, the variable must have the factor class. The "Variable Classes" section in Chapter 3 explains how to check the class of a variable and change it if necessary.

To create a bar chart from a previously saved one-dimensional table object (see "Frequency Tables" in Chapter 6), use the barplot function:

```
> barplot(tableobject)
```

For a horizontal bar chart like the one shown in Figure 8-7, set the horiz argument to T. This works with both the plot and barplot functions:

```
> plot(people2$Eye.Color, horiz=T)
```

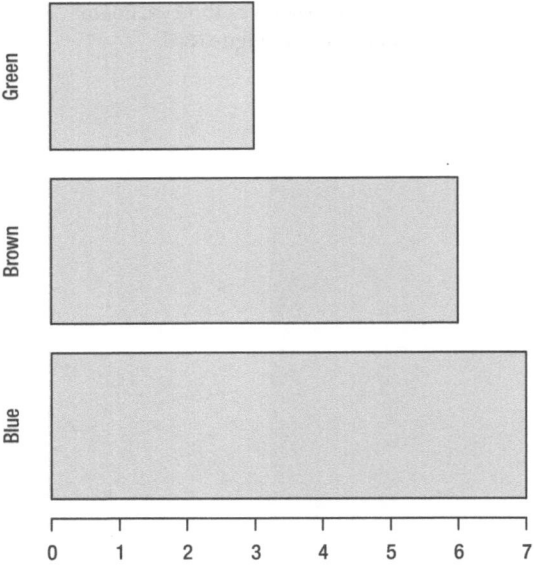

Figure 8-7. *Horizontal bar chart of the* Eye.Color *variable from the* people2 *dataset, created by setting* horiz=T

The barplot function can also create a bar chart for two categorical variables, known as a multiple bar chart or a clustered bar chart. There are two styles of multiple of bar chart, as shown in Figure 8-8.

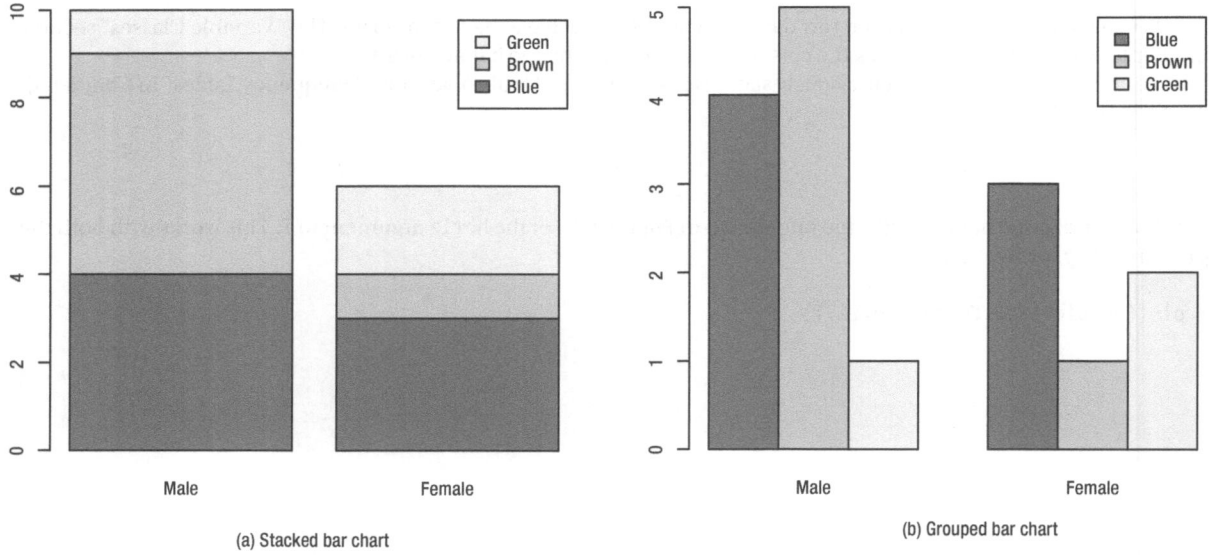

Figure 8-8. *Two styles of multiple bar chart of* Eye.Color *by* Sex

The first is the stacked style, which you can create from a two-dimensional table object:

```
> barplot(sexeyetable, legend.text=T)
```

The second style is a grouped bar chart, which displays the categories side-by-side. Create this style by setting the beside argument to T:

```
> barplot(sexeyetable, beside=T, legend.text=T)
```

To create multiple bar charts from raw data, first create a two-dimensional table object, as explained in the "Frequency Tables" section in Chapter 6. Alternatively, you can nest the table function inside the barplot function:

```
> barplot(table(people2$Eye.Color, people2$Sex), legend.text=T)
```

Pie Charts

The pie chart is a plot for a single categorical variable and is an alternative to the bar chart. It displays the number of observations in each category as a portion of the total number of observations.

You can create a pie chart with the pie function. If you have previously created a one-dimensional table object (see "Frequency Tables" in Chapter 6), you can use the function directly:

```
> pie(tableobject)
```

To create a pie chart from raw data, nest the table function inside the pie function as shown here. The result is given in Figure 8-9.

```
> pie(table(people2$Eye.Color))
```

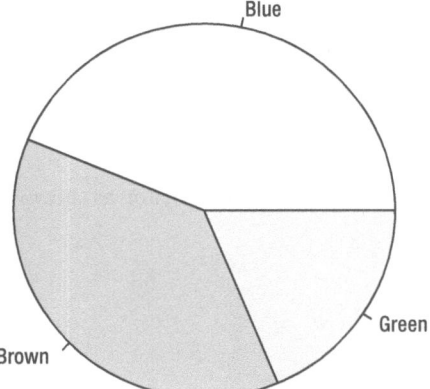

Figure 8-9. *Pie chart of the Eye.Color variable*

If your variable has missing data and you want this to appear as an additional section in the pie chart, set the useNA argument to "ifany":

```
> pie(table(dataset$variable, useNA="ifany"))
```

Scatter Plots

A scatter plot is a plot for two continuous variables, which allows you to examine the relationship between them.

You can create a scatter plot with the plot function, by giving two numeric variables as input. The first variable is displayed on the vertical axis and the second variable on the horizontal axis. For example, to plot Volume against Girth for the trees dataset, use the command:

```
> plot(Volume~Girth, trees)
```

The output is shown in Figure 8-10.

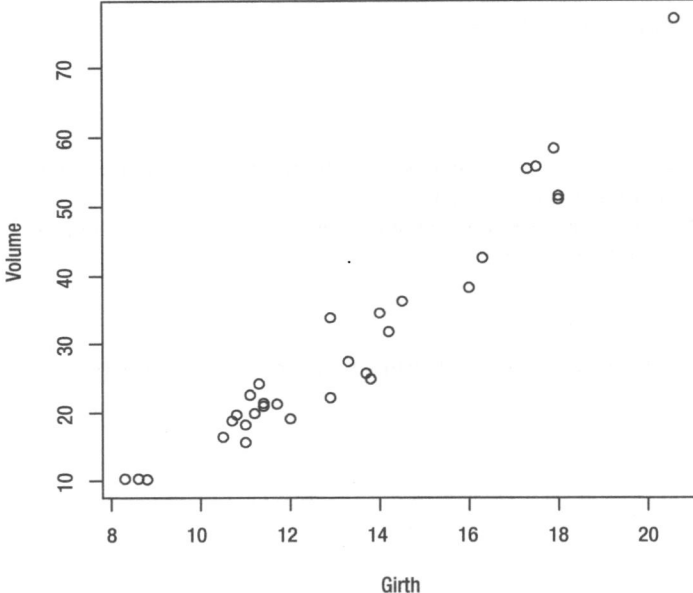

Figure 8-10. *Scatter plot of Volume against Girth, for the trees dataset*

To add a line of best fit (linear regression line), use the abline function directly after the plot function as shown here. Figure 8-11 shows the result.

```
> plot(Volume~Girth, trees)
> abline(coef(lm(Volume~Girth, trees)))
```

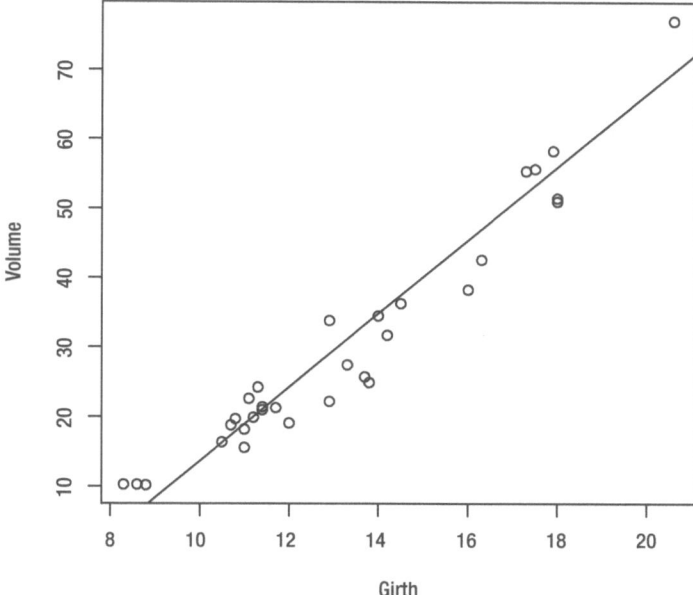

Figure 8-11. *Scatter plot of* Volume *against* Girth, *for the* trees *dataset, with line of best fit superimposed*

You will learn more about the abline function in Chapter 9 under "Adding Additional Straight Lines," and the lm and coef functions in Chapter 11.

If it is meaningful for your data, you can join the data points with lines by setting the type argument to "l". To create a plot with both symbols and lines, set it to "b". The effect is shown in Figure 8-12.

```
> plot(Volume~Girth, trees, type="l")
```

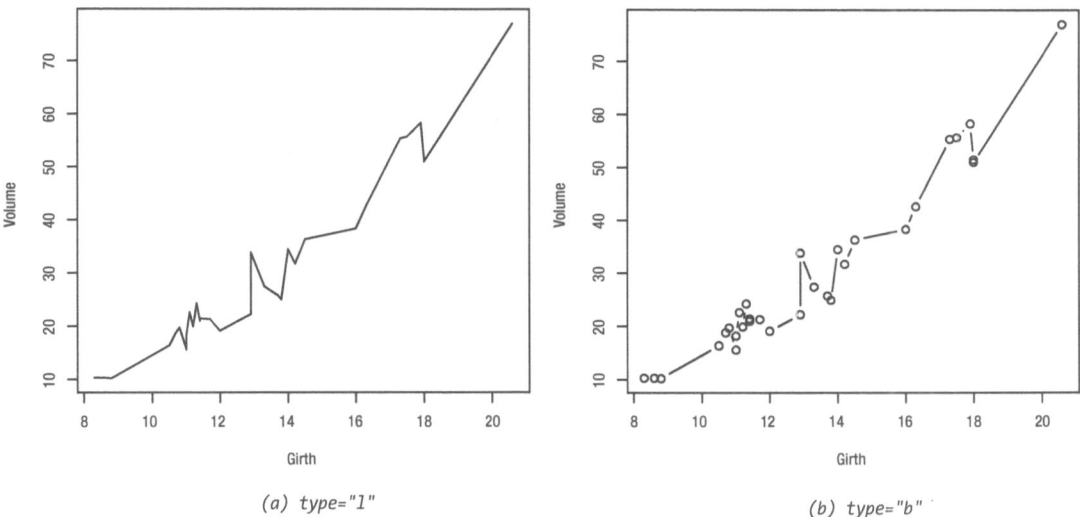

Figure 8-12. *Scatter plots with (a) lines and (b) both lines and symbols*

Scatterplot Matrices

A scatterplot matrix is a collection of scatter plots showing the relationship between each pair in a set of variables. It allows you to examine the correlation structure of a dataset.

To create a scatterplot matrix of the variables in a dataset, use the pairs function. For example, to create a scatterplot matrix for the iris dataset, use the command:

```
> pairs(iris)
```

The output is shown in Figure 8-13.

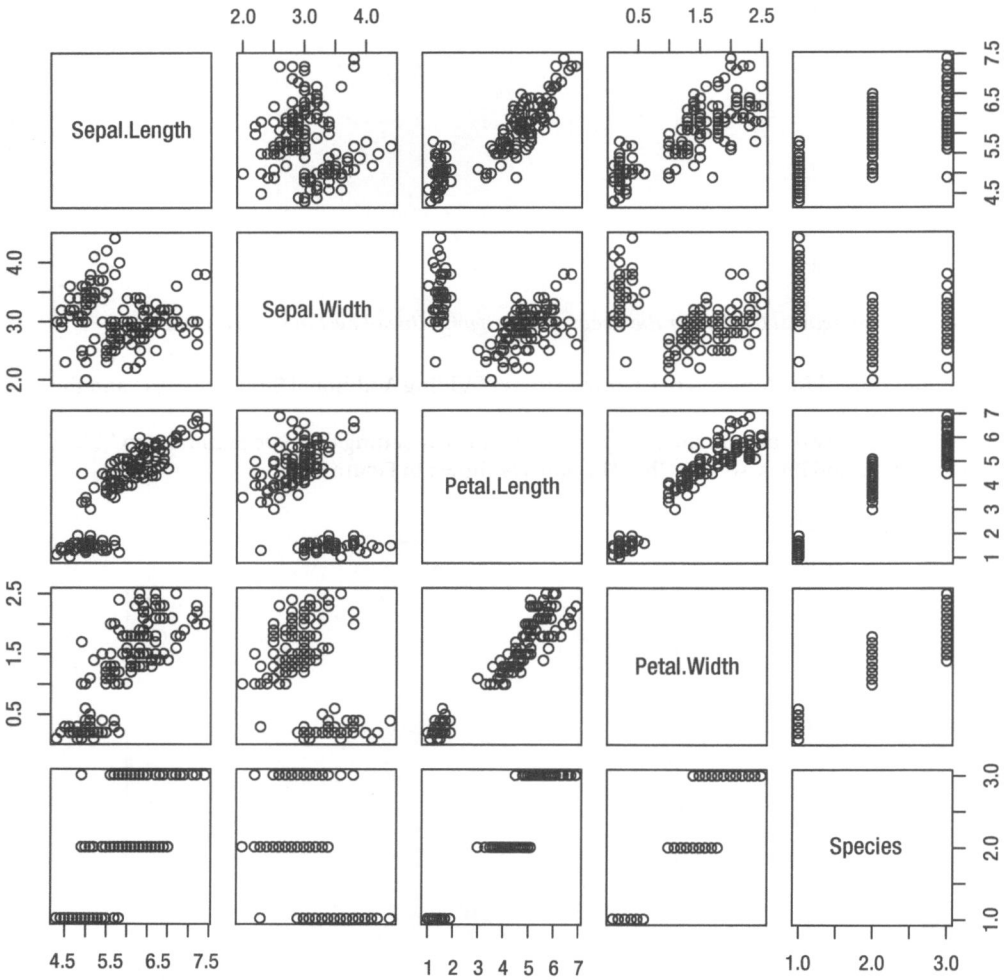

Figure 8-13. *Scatterplot matrix for the iris dataset*

You can also select a subset of variables to include in the matrix. For example, to include just the Sepal.Length, Sepal.Width and Petal.Length variables, use the command:

```
> pairs(~Sepal.Length+Sepal.Width+Petal.Length, iris)
```

Alternatively, you can use bracket notation or the subset function to select or exclude variables from a dataset, as explained in Chapter 1 under "Data Frames" and Chapter 3 under "Selecting a Subset of the Data," respectively.

Box Plots

A box plot (or box-and-whisker plot) presents summary statistics for a continuous variable in a graphical form. Usually, a categorical variable is used to group the observations, so that the plot summarizes the distribution for each category. This helps you to understand the relationship between a categorical and a continuous variable.

You can create a box plot with the boxplot function. For example, the following command creates a box plot of Sepal.Length grouped by Species for the iris dataset. Figure 8-14 shows the result.

```
> boxplot(Sepal.Length~Species, iris)
```

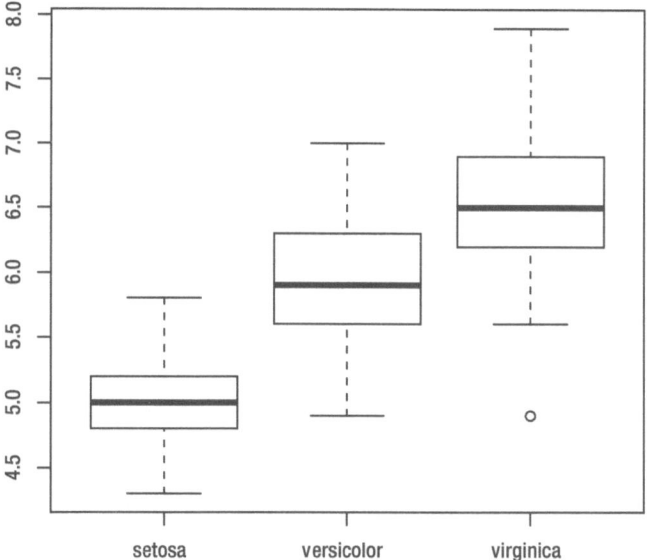

Figure 8-14. Box plot of Sepal.Length grouped by Species, for the iris dataset

When interpreting a box plot, recall that the thick line inside the box shows the group median and the boundaries of the box shows the interquartile range. The whiskers show the range of the data, excluding outliers, which are represented by small circles.

To create a horizontal box plot as shown in Figure 8-15, set the horizontal argument to ⁻:

```
> boxplot(Sepal.Length~Species, iris, horizontal=T)
```

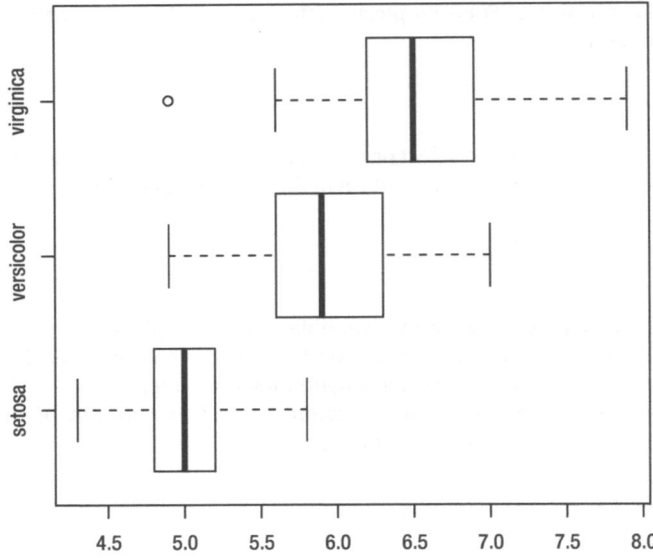

Figure 8-15. *Horizontal box plot of* Sepal.Length *grouped by* Species, *for the* iris *dataset*

You can also create a single box plot for a continuous variable (without any grouping):

```
> boxplot(iris$Sepal.Length)
```

By default, the whiskers extend to a maximum of 1.5 times the interquartile range of the data, with any values beyond this is shown as outliers. If you want the whiskers to extend to the minimum and maximum values, set the range argument to 0:

```
> boxplot(Sepal.Length~Species, iris, range=0)
```

Figure 8-16 shows the results.

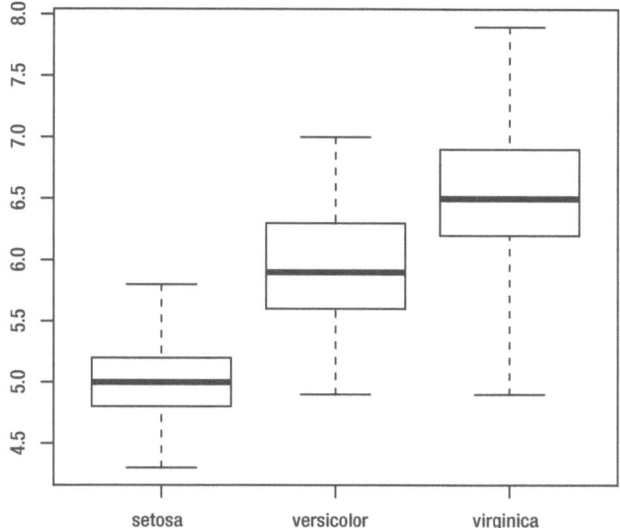

Figure 8-16. *Box plot of* Sepal.Length *grouped by* Species, *with whiskers showing the full range of the data*

Plotting a Function

You can plot a mathematical function such as $y=t^3$ with the curve function. Express the mathematical function in terms of x, as shown here. Figure 8-17 shows the result.

```
> curve(x^3)
```

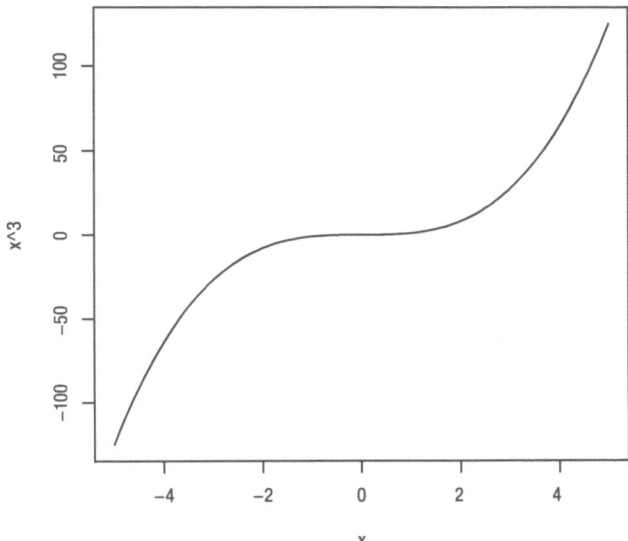

Figure 8-17. *Plots of the function* $y=x^3$, *created with the* curve *function*

■ **Note** To display the most relevant part of the curve, you may need to adjust the axes, as explained under "Axes" in Chapter 9.

To plot a curve of the probability density function (pdf) for a standard probability distribution, use the relevant density function from Table 7-1 (see Chapter 7). For example, to plot the pdf of the standard normal distribution, use the command:

```
> curve(dnorm(x))
```

The results are shown in Figure 8-18.

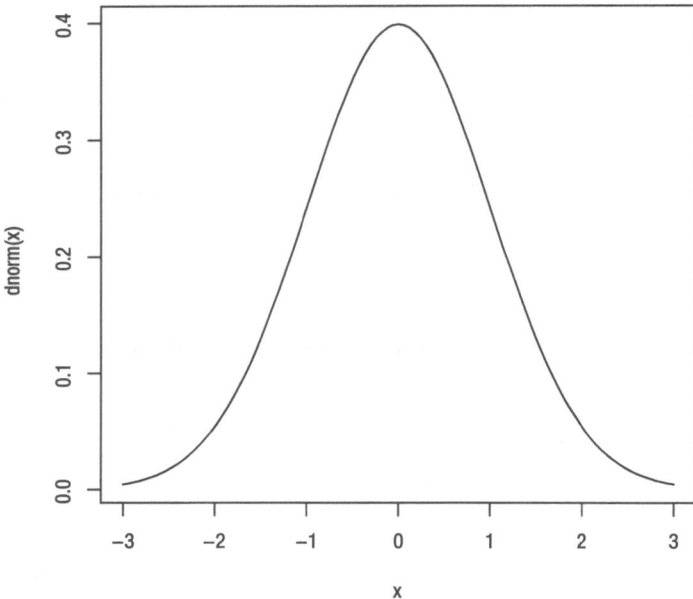

Figure 8-18. *Plot of the density function for the normal distribution*

Exporting and Saving Plots

Before exporting or saving a plot, resize the graphics device window so that the plot has the required dimensions.

For Windows users, the simplest way to export a plot is to right-click on the plot and select Copy as bitmap. The image is copied to the clipboard, so that you can paste it into a Microsoft Word or Powerpoint file or a graphics package.

Alternatively, R allows you to save the image to a variety of file types including png, jpeg, bmp, pdf, and eps. With the graphics device as the active window, select Save As from the File menu. You will be given a number of options for saving the image.

Mac users can copy the image to the clipboard by selecting Edit then Copy, or save the image as a pdf file by selecting File, then Save.

Linux users can save the current plot by entering the command:

```
> savePlot("/home/Username/folder/filename.png", type="png")
```

Other possible file types are bmp, jpeg, and tiff.

Summary

This chapter looked at how to create the most popular plot types for continuous and categorical data. This included plots that allow you to look at the distribution of a single variable as well as plots that help you to examine the relationship between two or more variables. You also learned how to save your plot to an image file or paste it into another program.

This table summarizes the main commands covered.

Plot Type	Command
Basic plot	plot(*dataset$variable*)
Line plot	plot(*dataset$variable*, type="l")
Histogram	hist(*dataset$variable*)
Normal probability plot	qqnorm(*dataset$variable*)
Stem-and-leaf plot	stem(*dataset$variable*)
Bar chart	plot(*dataset$factor1*)
	barplot(*tableobject1D*)
Stacked bar chart	barplot(*tableobject2D*)
Grouped bar chart	barplot(*tableobject2D*, beside=T)
Pie chart	pie(*tableobject1D*)
	pie(table(*dataset$factor1*))
Scatter plot	plot(*yvar~xvar*, *dataset*)
Scatterplot matrix	pairs(*dataset*)
Single box plot	boxplot(*dataset$variable*)
Grouped box plot	boxplot(*variable~factor1*, *dataset*)
Function plot	curve($f(x)$)

In the next chapter, you will learn how to customize the appearance of your plots so that they look more presentable.

CHAPTER 9

■ ■ ■

Customizing Your Plots

In the previous chapter, you learned how to create the most popular types of plots. This chapter explains how you can customize their appearance. Good use of titles, labels, and legends is essential if you want your plots to be useful and informative to others, and carefully selected colors and styles will help to create a more polished look.

You will learn how to:

- add titles and axis labels
- adjust the axes
- specify colors
- change the plotting line style
- change the plotting symbol
- adjust the shading style of shaded areas
- add items such as straight lines, function curves, text, grids, and arrows
- overlay several groups of data onto one plot
- add a legend
- display multiple plots in one image
- change the default plot settings

Most changes to the appearance of a plot are made by giving additional arguments to the plotting function. In this chapter, the plot function is used to illustrate. However, most of the arguments can also be used with other plotting functions such as hist and boxplot.

This chapter uses the trees and iris datasets, which are included with R, the CIAdata dataset (created in Chapter 4), and the people2 and fiveyearreport datasets, which are available with the downloads for this book.

Titles and Labels

Whenever you create a plot, R adds default axis labels and sometimes a default title, too. You can overwrite these with your own titles and axis labels.

You can add a title to your plot with the main argument:

```
> plot(dataset$variable, main="This is the title")
```

119

To add a subtitle to be displayed underneath the plot, use the sub argument:

```
> plot(dataset$variable, sub="This is the subtitle")
```

If you have a very long title or subtitle that should be displayed over two of more lines, use the escape sequence \n to tell R where the line break should be. The command below adds a two-line title to a plot of tree heights, as shown in Figure 9-1.

```
> plot(trees$Height, main="This is the first line of the title\n and this is the second line")
```

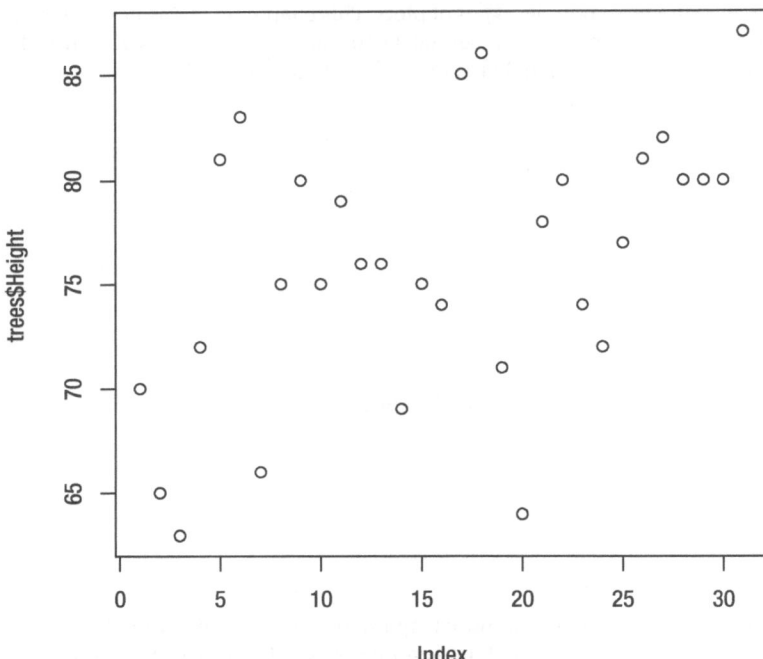

Figure 9-1. *Plot with two-line title*

To change the *x* and *y* axis labels, use the xlab and ylab arguments:

```
> plot(dataset$variable, xlab="The x axis label", ylab="The y axis label")
```

You can also add additional arguments to control the appearance of the text in the title, subtitle, axis labels, and axis units. Table 9-1 gives a list of these.

Table 9-1. *Additional arguments for customizing the font of titles and labels*

Aspect		Argument	Possible values
Font family	All text	`family="serif"`	`sans`, `serif`, `mono`
Font type	Title	`font.main=2`	1 (normal)
	Subtitle	`font.sub=2`	2 (bold)
	Axis labels	`font.lab=2`	3 (italic)
	Axis units	`font.axis=2`	4 (bold & italic)
Font size	Title	`cex.main=2`	Relative to default, e.g. 2 is
	Subtitle	`cex.sub=2`	twice normal size
	Axis labels	`cex.lab=2`	
	Axis units	`cex.axis=2`	
Font color	Title	`col.main="red"`	See the "Colors" section
	Subtitle	`col.sub="red"`	later in this chapter for
	Axis labels	`col.lab="red"`	details of how to specify
	Axis units	`col.axis="red"`	colors

For example, this command adds a title, sets the font family to serif, and the title font to be gray, italic, and three times the usual size. The result is shown in Figure 9-2.

```
> plot(trees$Height, main="This is the title", family="serif", col.main="grey80", cex.main=3, font.main=3)
```

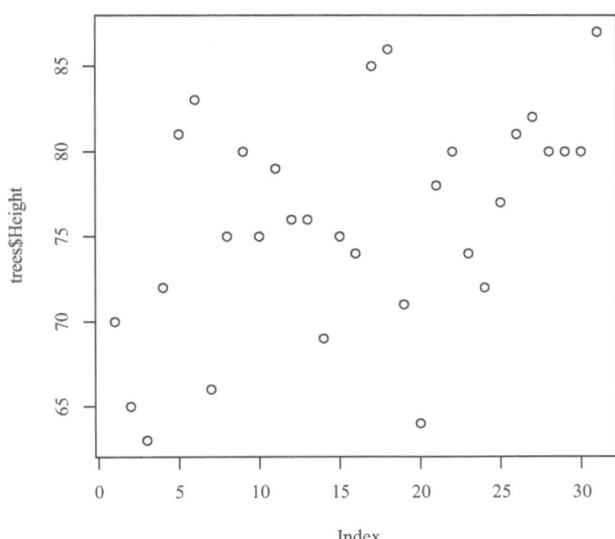

Figure 9-2. *Plot with serif font and large gray italic title*

For plots that include a categorical variable such as the bar chart, pie chart, and box plot, you can also customize the category labels. If you plan to create several plots, then it is easier to change the level names for the factor variables in your dataset, as explained in Chapter 3 under "Working with Factor Variables". This means you won't have to modify the labels every time you plot the variable. However, it is also possible to modify the labels at the time of plotting.

For pie charts, use the `labels` argument:

```
> pie(table(dataset$variable), labels=c("Label1", "Label2", "Label3"))
```

For bar charts, use `names.arg`:

```
> plot(dataset$variable, names.arg=c("Label1", "Label2", "Label3"))
```

For box plots, use `names`:

```
> boxplot(variable~factor, dataset, names=c("Label1", "Label2", "Label3"))
```

Axes

When plotting continuous variables, R automatically selects sensible axis limits for your data. However, if you want to change them, you can do so with the `xlim` and `ylim` arguments. For example, to change the axis limits to 0–30 for the horizontal axis and 0–100 for the vertical axis, use the command:

```
> plot(Volume~Girth, trees, xlim=c(0, 30), ylim=c(0, 100))
```

Figure 9-3 shows the results.

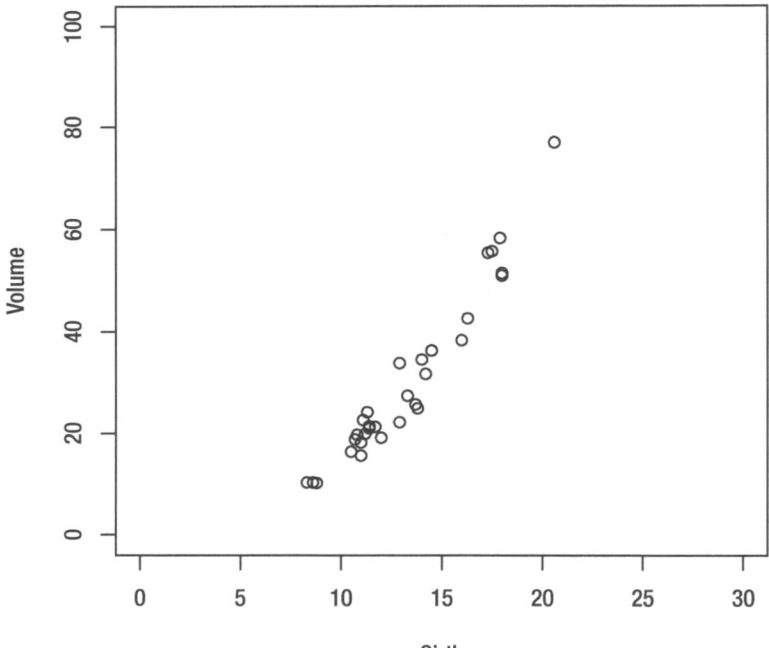

Figure 9-3. *Plot with adjusted axis ranges*

To rotate the axis numbers for the vertical axis so that they are in upright rather than in line with the vertical axis, set the las argument to 1. The result is shown in Figure 9-4.

```
> plot(Volume~Girth, trees, las=1)
```

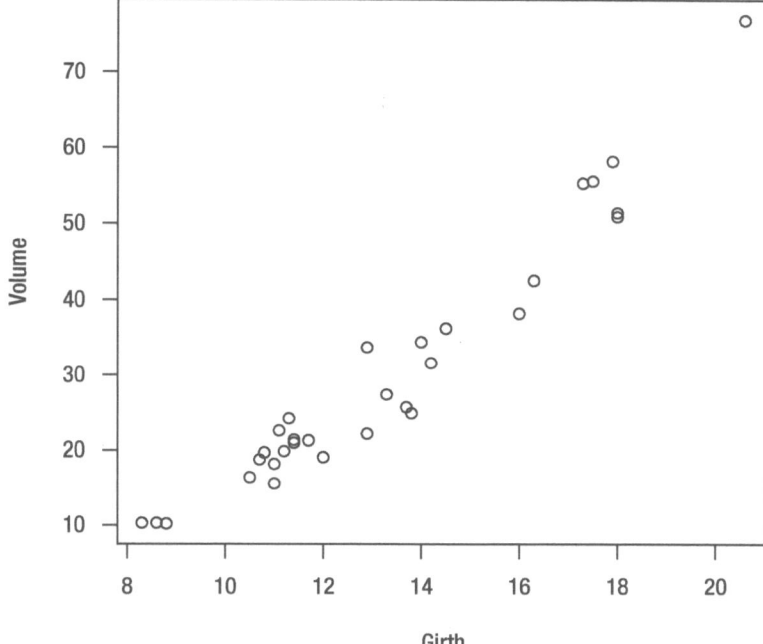

Figure 9-4. *Plot with rotated axis numbers*

Colors

R allows you to change the color of every component of a plot. The most important argument is the col argument, which allows you to change the plotting color. This is the color of the lines or symbols that are used to represent the data values. For histograms, bar charts, pie charts, and box plots it is the color of the area enclosed by the shapes.

There are three ways to specify colors. The first is to give the name of the color:

```
> plot(dataset$variable, col="red")
```

The colors function displays a list of recognized color names (the British-English spelling colours also works):

```
> colors()
```

The second way to specify colors is with a number between 1 and 8, which correspond to the set of basic colors shown in Table 9-2.

Table 9-2. *Basic plotting colors; note that these may vary for some platforms; enter* `palette()` *to view the colors for your platform*

Value	Name
1	black
2	red
3	green3
4	blue
5	cyan
6	magneta
7	yellow
8	gray

For example, to specify red as the plotting color, use the command:

```
> plot(dataset$variable, col=2)
```

Finally, if you require more precise control over the appearance of the plot then you can specify the color using the hexadecimal RGB format. This command changes the plotting color to red using the hexadecimal color notation:

```
> plot(dataset$variable, col="#FF0000")
```

HEXADECIMAL COLOR NOTATION

Hexadecimal color notation (sometimes called hex colors or html colors) allows you fine control over the colors in your graphics.

In the hexadecimal color notation, each color is regarded as unique combination of the three primary colors (red, green, and blue), each with a strength between 00 and FF. This combination is written in the format #RRGGBB. For example, bright red is written as #FF0000, white is written #FFFFFF, and black is written #000000.

To identify the hexadecimal code of a color in an image, use a graphics package or a website such as `http://imagecolorpicker.com/`. This is useful if you want to make elements of a plot match your slides or company logo.

To design a color scheme of hexadecimal colors from scratch, try `http://colorschemedesigner.com/`.

As well as specifying a single plotting color, you can give a list of colors for R to use. This is particularly useful for pie charts and bar charts, where each category is given a different color from the list:

```
> plot(dataset$variable, col=c("red", "blue", "green"))
```

The numeric notation is useful for this purpose:

```
> plot(dataset$variable, col=1:8)
```

There are also functions such as rainbow, which help to easily create a visually appealing set of colors. This command creates five colors, evenly spaced along the spectrum:

```
> plot(dataset$variable, col=rainbow(5))
```

Other color scheme functions include heat.colors, terrain.colors, topo.colors, and cm.colors.
You can change the color of the other elements of the plot in a similar way, using the arguments given in Table 9-3.

Table 9-3. *Arguments for changing the color of plot component (*the background color must be changed with the* par *function)*

Component	Argument
Plotting symbol, line, or area	col
Foreground (axis, tick marks, and other elements)	fg
Background	bg*
Title text	col.main
Subtitle text	col.sub
Axis label text	col.lab
Axis numbers	col.axis

Note that the background color cannot be changed from within the plotting function but must be changed with the par function:

```
> par(bg="red")
> plot(dataset$variable)
```

Changes made with the par function apply to all plots for the remainder of the session, or until overwritten.

■ **Note** For more details on using the par function, see "Changing the Default Plot Settings" later in this chapter, or enter help(par).

Plotting Symbols

This section applies only to plots where symbols are used to represent the data values, such as scatter plots.
You can change the plotting symbol with the pch argument. The numbers 1 to 25 correspond to the symbols given in Table 9-4.

```
> plot(dataset$variable, pch=5)
```

Table 9-4. *Plotting symbols for use with the pch argument*

Number	Symbol	Number	Symbol	Number	Symbol
1	○	10	⊕	19	●
2	△	11	✕	20	●
3	＋	12	⊞	21	○
4	✕	13	⊠	22	□
5	◇	14	◩	23	◇
6	▽	15	■	24	△
7	⊠	16	●	25	▽
8	✳	17	▲		
9	⊕	18	◆		

Symbol numbers 21 to 25 allow you to specify different colors for the symbol border and fill. The border color is specified with the col argument and the fill color with the bg argument:

```
> plot(dataset$variable, pch=21, col="red", bg="blue")
```

■ **Note** For more information on plotting symbols, enter the command help(points).

Alternatively, you can select any single keyboard character to use as the plotting symbol by placing it between quotation marks, as shown here for the dollar symbol:

```
> plot(dataset$variable, pch="$")
```

You can adjust the size of the plotting symbol with the cex argument. The size is specified relative to the normal size. For example, to make the plotting symbols 5 times their normal size, use the command:

```
> plot(dataset$variable, cex=5)
```

Plotting Lines

This section applies only to plots with lines, such as scatter plots and basic plots created with the plot function and where the type argument is set to an option that uses lines such as "l" or "b". It also applies to functions that add additional lines to existing plots, such as abline, curve, segments and lines, covered later in this chapter in the sections "Adding Items to Plots" and "Overlaying Plots."

You can change the line type with the lty argument. The numbers 1 to 6 correspond to the line types given in Table 9-5.

Table 9-5. *Plotting line types*

Number	Name	Line Style
1	solid	————————————————
2	dashed	— — — — — — — — — — — —
3	dotted	··································
4	dotdash	·—·—·—·—·—·—·—·—
5	longdash	—— —— —— —— —— —— ——
6	twodash	·—·——·—·——·—·——

For example, you can change the line type to dashed:

```
> plot(dataset$variable, type="l", lty=2)
```

You can also select a line type by name:

```
> plot(dataset$variable, type="l", lty="dashed")
```

You can adjust the thickness of the line with the lwd argument. The line thickness is specified relative to the standard line thickness. For example, to make the line three times its usual thickness, use the command:

```
> plot(dataset$variable, type="l", lwd=3)
```

Shaded Areas

This section applies only to plots with shaded areas such as bar charts, pie charts, and histograms.

The *density* of a shaded area is the number of lines per inch used to shade. You can change the density of the shaded areas with the density argument and the angle of the shading with the angle argument:

```
> barplot(tableobject, density=20, angle=30)
```

As a guideline, density values between 5 and 80 give discernible variation, while a value of 150 is indistinguishable from solid color.

You can also give a list of densities and R will shade each section of a pie or bar chart with a different density from the list. Figure 9-5 shows the result of the following:

```
> pie(table(people2$Eye.Color), density=c(10, 20, 40))
```

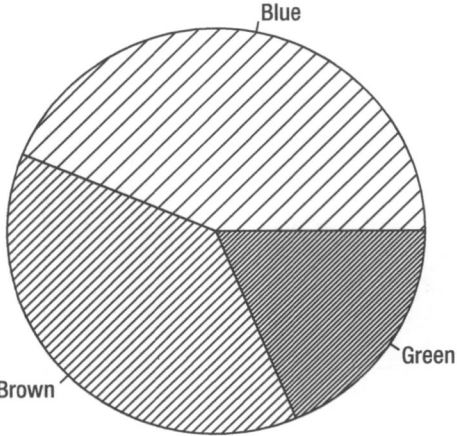

Figure 9-5. *Pie chart with a different density of shading for each category*

Adding Items to Plots

This section introduces some functions that add extra items to plots. You can use them with any type of plot. First, create the plot using plot or another plotting function. While the plot is still displayed in the graphics device, enter the relevant command from this section. The item is added to the current plot.

Adding Straight Lines

You can add straight lines to your plot with the abline function.

To add a vertical line at x=5 use the command:

```
> abline(v=5)
```

To add a horizontal line at y=2 use:

```
> abline(h=2)
```

To add a diagonal line with intercept 2 and slope 3 (i.e., the line y=2+3x) use the command:

```
> abline(a=2, b=3)
```

To draw a line segment (a line that extends from one point to another), use the segments function. For example, the command below draws a straight line from coordinate (12, 20) to coordinate (18, 55):

```
> segments(12,20,18,55)
```

The result is shown in Figure 9-6.

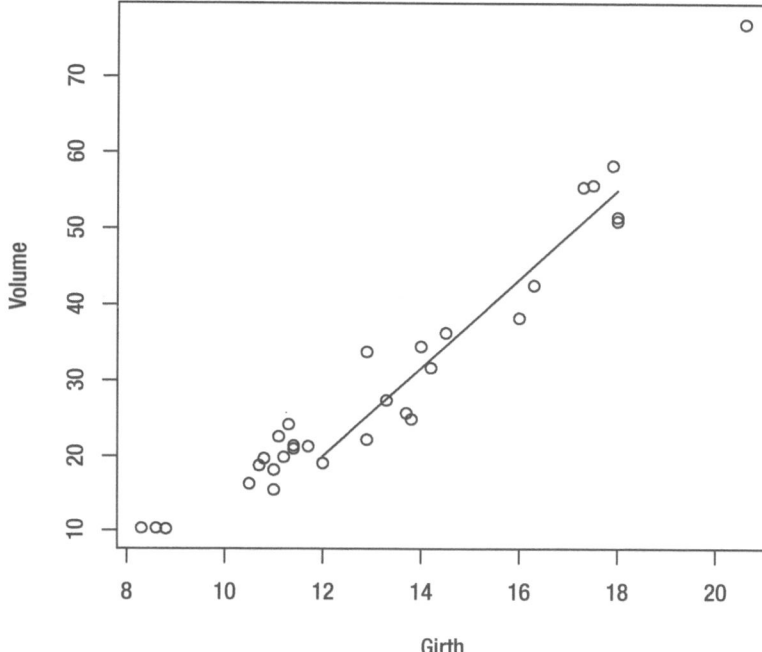

Figure 9-6. *Plot with added line segment*

You can change the line color, type, and thickness with the col, lty, and lwd arguments, as explained earlier in the chapter.

Adding a Mathematical Function Curve

To add a mathematical function curve to your plot, use the curve function (introduced in Chapter 8 under "Plotting a Function"). By default the curve function creates a new plot. To superimpose the curve over the current plot, set the add argument to T:

```
> curve(x^2, add=T)
```

Adding Labels and Text

You can add text to your plot with the text function. For example, to add the text '*Text String*' centered at coordinates (3, 4), use the command:

```
> text(3, 4, "Text String")
```

> ■ **Note**   If you need to know the coordinates of a given location on your plot, the locator function can tell you them. While your plot is displayed in the graphics device, enter the command:
>
> > locator(1)
>
> R will then wait for you to select a location in the graphics device with your mouse pointer before returning the coordinates of the selected location.

You can adjust the appearance of the text with family, font, cex, and col arguments. For example, to add a piece of red, italic text which is twice the default size, use the command:

```
> text(3, 4, "Text String", col="red", font=3, cex=2)
```

The text function is useful if you want to add labels to all of the points in a scatter plot using text taken from a third variable or from the row names of your dataset. For example, these commands create a scatter plot of per capita GDP against urban population, where each observation is labelled with the country name (using the CIAdata dataset shown in Chapter 4 under "Appending Rows"):

```
> plot(pcGDP~urban, CIAdata, xlim=c(50, 100), ylim=c(0,40000))
> text(pcGDP~urban, CIAdata, CIAdata$country, pos=1)
```

Figure 9-7 shows the results.

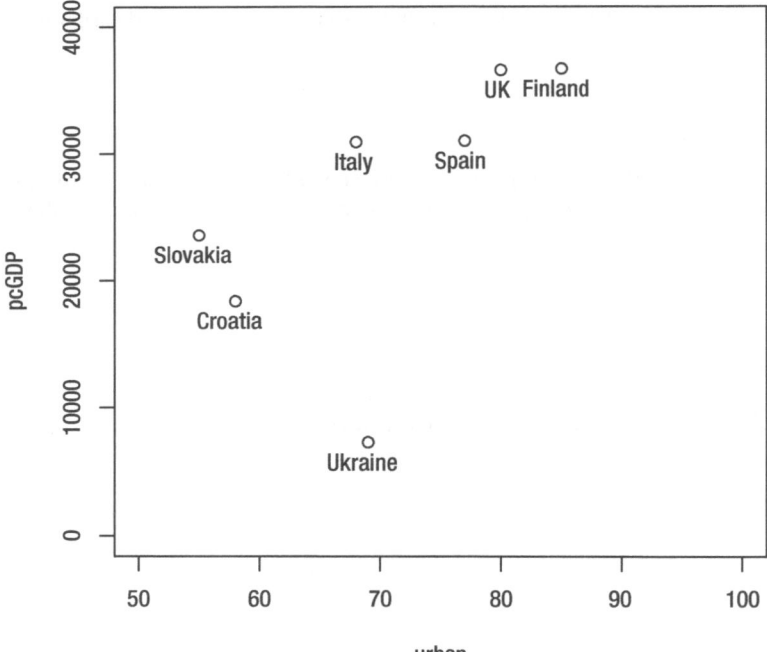

Figure 9-7. *Results of using the text function to add labels*

The pos argument tells R where to place the labels in relation to the coordinates: 1 is below; 2 is to the left, 3 is above; 4 is to the right. You can also use the offset argument to adjust the distance (relative to the character width) between the coordinate and the label. The command below creates the result shown in Figure 9-8.

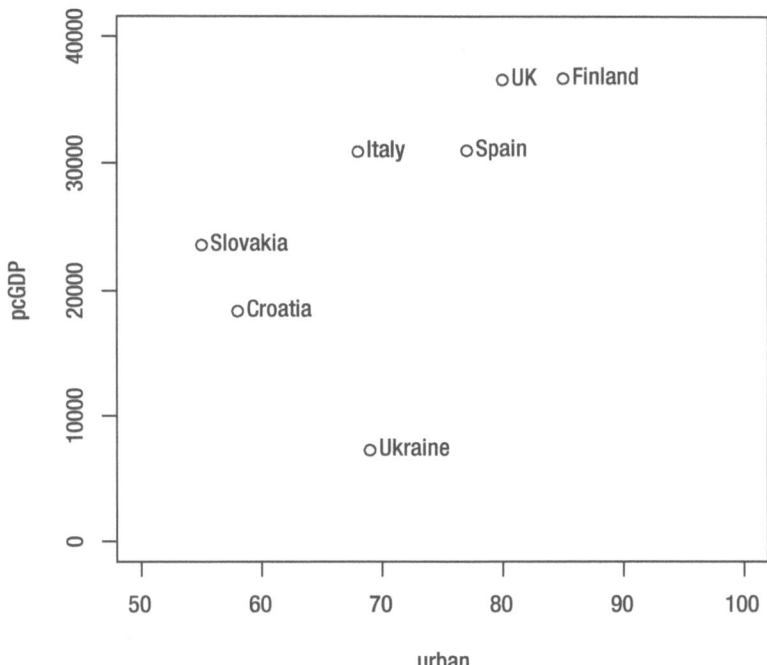

Figure 9-8. *Plot with text positioned to the right of data points*

```
> text(pcGDP~urban, CIAdata, CIAdata$country, pos=4, offset=0.3)
```

If you just want to label a few specific data points (such as any outliers), you can use the identify function after the plotting function:

```
> plot(pcGDP~urban, CIAdata, xlim=c(50, 100), ylim=c(0,40000))
> identify(CIAdata$urban, CIAdata$pcGDP, label=CIAdata$country)
```

After you have entered the command, R will allow you to select points on the plot using your mouse pointer, and will label any points that you select. Once you have selected all of the points that you want to label, press the Esc key.

Adding a Grid

To add a grid to a plot (as shown in Figure 9-9), use the grid function:

```
> grid()
```

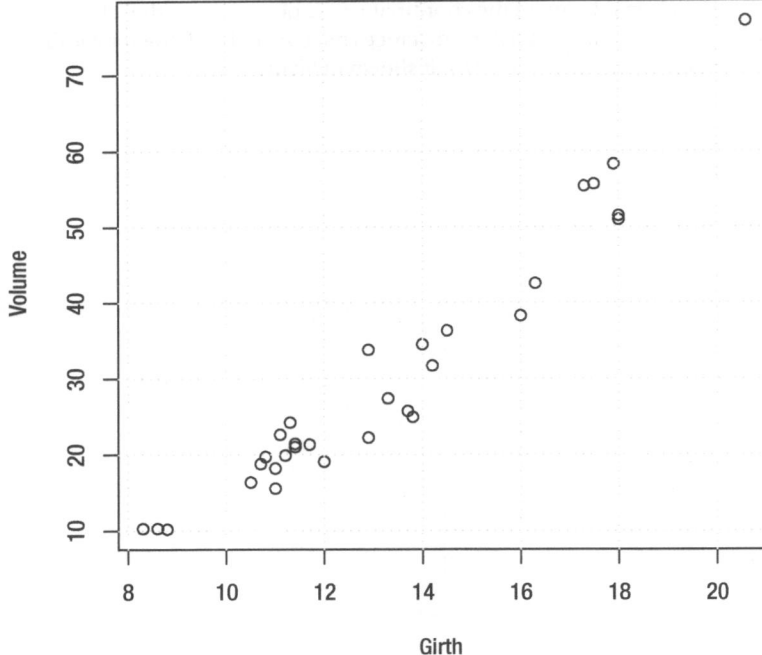

Figure 9-9. *Plot with grid lines*

By default, the grid lines are aligned with the axis tick marks. Alternatively, you can specify how many grid lines to display on each axis:

```
> grid(3, 3)
```

To add horizontal grid lines only (as shown in Figure 9-10), use the command:

```
> grid(nx=NA, ny=NULL)
```

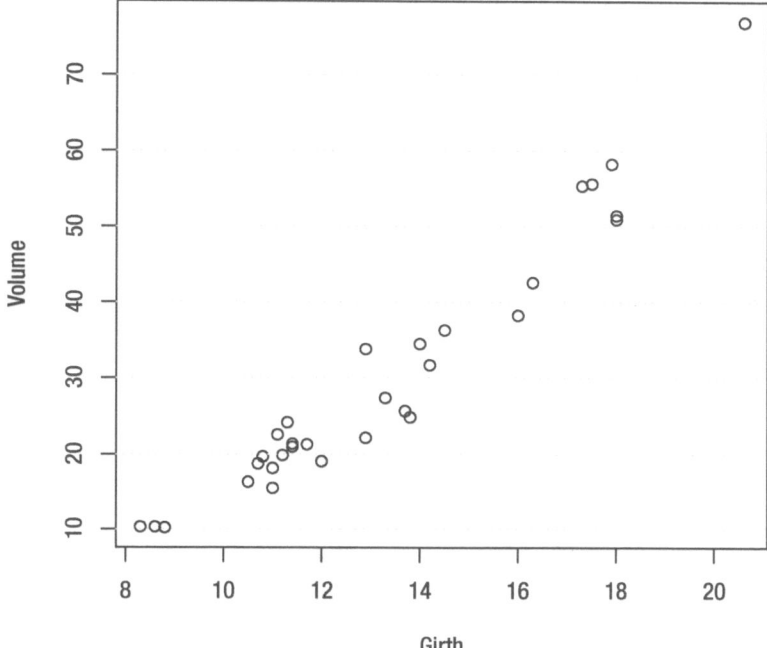

Figure 9-10. *Plot with horizontal grid lines*

For vertical grid lines only, use:

```
> grid(ny=NA)
```

By default, R uses grey dotted lines for the grid, but you can adjust the line style with the col, lty, and lwd arguments, as explained earlier in this chapter.

Adding Arrows

You can add an arrow to the current plot with the arrows function. For example, to draw an arrow from the point (10, 50) to the point (12.7, 35) use the command:

```
> arrows(10,50,12.7,35)
```

The result is shown in Figure 9-11.

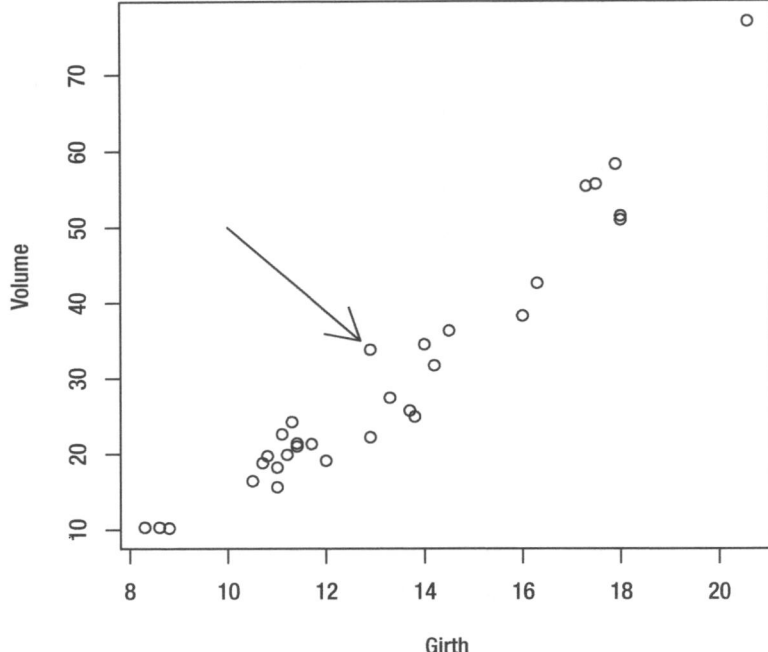

Figure 9-11. *Plot with arrow*

For a double-ended arrow, set the code argument to 3:

```
> arrows(10,50,12.7,35, code=3)
```

You can adjust the line style using the col, lty and lwd, as explained earlier in this chapter. You can specify the length of the arrow head (in inches) with the length argument, and the angle between the arrow head and arrow body (in degrees) with the angle argument. This command adds the arrow shown in Figure 9-12:

```
> arrows(10,50,12.7,35, angle=25, length=0.1)
```

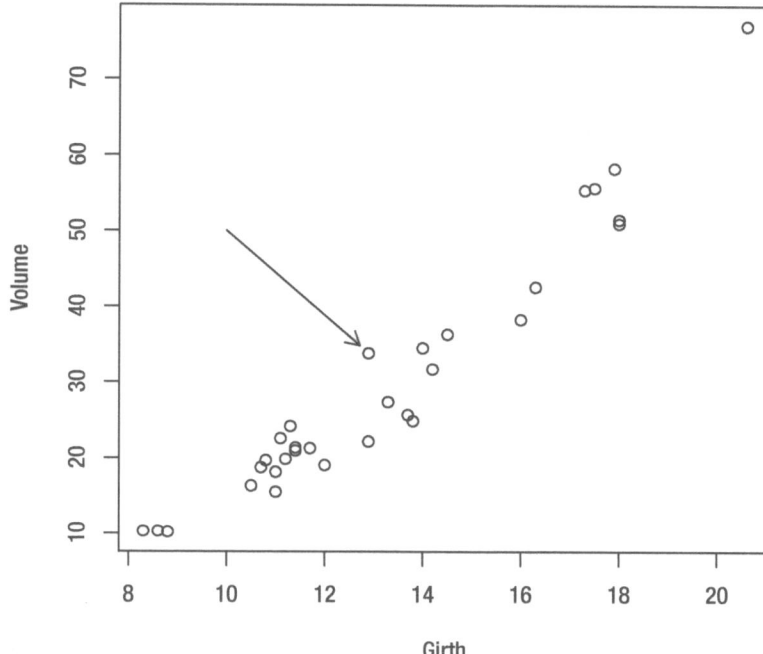

Figure 9-12. *Plot including arrow with adjusted arrow head*

Overlaying Plots

R has two functions called points and lines that allow you to superimpose one set of data values over another.

The points function is very similar to the plot function. However, instead of creating a new plot, it adds the data points to whichever plot is currently displayed in the graphics device. The lines function is very similar to the points function, except that it uses lines rather than symbols to plot the data (i.e., the default value for the type argument is "l").

These functions are useful if you want to plot two or more variables on the same axis (as demonstrated in Example 9-1) or use different plotting symbols to represent different categories (as shown in Example 9-2).

EXAMPLE 9-1. OVERLAY PLOT USING FIVEYEARREPORT DATA

Consider the fiveyearreport dataset shown in Figure 9-13, which gives the U.K. sales of three supermarket chains over a five-year period. Suppose that you want to display the sales data for Tesco, Sainsburys, and Morrisons in the same plot.

	Year	Tesco	Sainsburys	Morrisons
1	2007	35580	18518	12462
2	2008	37949	19287	12969
3	2009	41520	20383	14528
4	2010	42254	21421	15410
5	2011	44570	22943	16479

Figure 9-13. *fiveyearreport dataset (see Appendix C for more details)*

To create this plot, first plot the data for Tesco in the usual way, making sure to set the axis ranges so that they are wide enough to also accommodate the data for Sainsburys and Morrisons:

```
> plot(Tesco~Year, fiveyearreport, type="b", ylab="UK Sales (£M)", ylim=c(0, 50000))
```

Once the plot is displayed in the graphics device, you can add the data for Sainsburys and Morrisons using the lines function, which superimposes the data over the current plot. Use the pch and lty arguments to change the symbol and line type, so that the data for the different chains can be distinguished. Alternatively, you could use the col argument to give each chain a different color:

```
> lines(Sainsburys~Year, fiveyearreport, type="b", pch=2, lty=2)
> lines(Morrisons~Year, fiveyearreport, type="b", pch=3, lty=3)
```

Figure 9-14 shows the results. The plot will require a legend to identify the chains, which you can add with the legend function as explained in the "Adding a Legend" section.

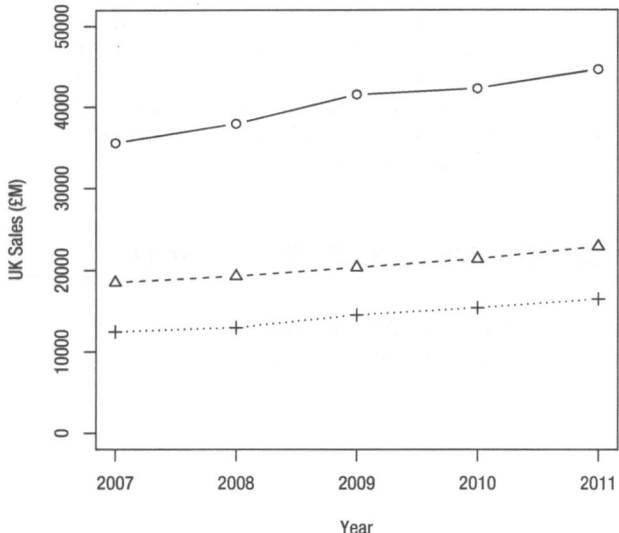

Figure 9-14. *Overlay plot for the fiveyearreport dataset*

EXAMPLE 9-2. OVERLAY PLOT USING IRIS DATA

Suppose that you want to create a scatter plot of sepal length against sepal width for the `iris` data, using a different plotting symbol to represent each iris species. To create this plot, you will need to plot the data for each species separately using the `points` function, overlaying them onto the same plot.

First plot the original (complete) data with the `plot` function, adding any labels or titles that you require. Set the `type` argument to `"n"` to prevent the data values from being added to the plot. This creates an empty axis, which is the right size to accommodate all of the data for all three species:

```
> plot(Sepal.Length~Sepal.Width, iris, type="n")
```

Next, plot the data for each species separately with the `points` function. Use the `subset` argument to select each species in turn. Use the `pch` argument to select a different plotting symbol for each species. Alternatively, you could use the `col` argument to give each species a different colored symbol:

```
> points(Sepal.Length~Sepal.Width, iris, subset=Species=="setosa", pch=10)
> points(Sepal.Length~Sepal.Width, iris, subset=Species=="versicolor", pch=16)
> points(Sepal.Length~Sepal.Width, iris, subset=Species=="virginica", pch=1)
```

The result is shown in Figure 9-15. The legend is added with the `legend` function, as explained in the next section.

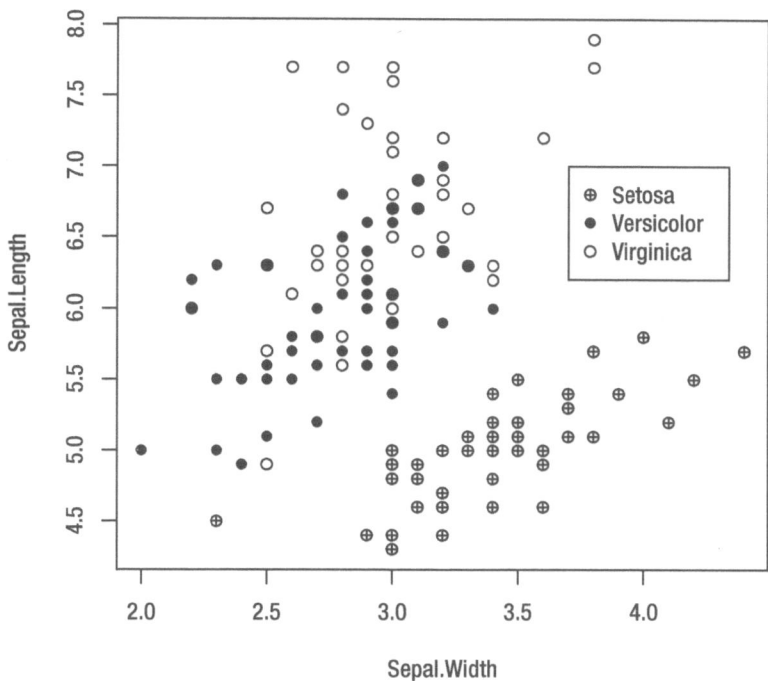

Figure 9-15. *Overlay plot for the `iris` data*

Adding a Legend

You can add a legend to your plot with the legend function. The function adds a legend to whichever plot is currently displayed in the graphics device.

The following command creates the legend shown in Figure 9-15:

```
> legend(3.7, 7, legend=c("Setosa", "Versicolor", "Virginica"), pch=c(10, 16, 1))
```

The first two arguments (3.7 and 7) are the *x* and *y* coordinates for the top left-hand corner of the legend. In addition to using coordinates, there are two other ways of specifying the position of the legend. The first is by using one of the location names: "top", "bottom", "left", "right", "center", "topright", "topleft", "bottomright", or "bottomleft". The following command positions the legend in the top right-hand corner of the plot:

```
> legend("topright", legend=c("Setosa", "Versicolor", "Virginica"), pch=c(10, 16, 1))
```

The second is by using locator(1). This allows you to manually select the position for the top left-hand corner of the legend with your mouse:

```
> legend(locator(1), legend=c("Setosa", "Versicolor", "Virginica"), pch=c(10, 16, 1))
```

The legend argument gives the labels to be displayed on the legend. The pch argument gives the plotting symbols that correspond to each of the labels.

You can substitute the pch argument with the lty, lwd, col, cex, fill, density, or angle arguments, according to what is relevant for your plot. You may need to use two or more of them.

Note that you don't need to use the legend function to add a legend to a bar chart, as the barplot function has a built-in legend option (see "Bar Charts" in Chapter 8).

EXAMPLE 9-3. LEGEND TYPES AND EFFECTS

The following three examples demonstrate how to create some different types of legend. The results are shown in Figure 9-16. Then we'll look at another example of how to display a legend over more than one column (shown in Figure 9-17).

Legend Showing Shaded Areas of Different Densities

To create a legend for shaded areas of different densities as shown in Figure 9-16a, use the command:

```
> legend(locator(1), legend=c("Label1", "Label2", "Label3"), density=c(10,20,40))
```

Legend Showing Different Line and Symbol Types

To create a legend for different line and symbol types as shown in Figure 9-16b, use the command:

```
> legend(locator(1), legend=c("Label1", "Label2", "Label3"), lty=1:3, pch=1:3)
```

Legend with Lines of Different Colors and Types

To create a legend for lines of different types and colors as shown in Figure 9-16c, use the command:

```
> legend(locator(1), legend=c("Label1", "Label2", "Label3"), col=c("black", "grey40",
        "grey70"), lty=1:3)
```

(a)   *(b)*   *(c)*

Figure 9-16. *Different types of legend*

Legend That Extends over Multiple Columns

To display a legend over more than one column, use the ncol argument to specify the number of columns. This command creates a legend with two columns, as shown in Figure 9-17:

```
> legend(x, y, legend=c("Label1", "Label2", "Label3"), lty=c(1,2,3), ncol=2)
```

```
┌─────────────────────────────────────┐
│ ────  Label1    ······  Label3       │
│ - - -  Label2                         │
└─────────────────────────────────────┘
```

Figure 9-17. *Legend with two columns*

Multiple Plots in the Plotting Area

As well as creating images of single plots, R allows you to create an image composed of several smaller plots arranged in a grid formation.

To set up the graphics device to display multiple plots, use the command:

```
> par(mfrow=c(R,C))
```

where R and C are the number of row and columns in the grid. Any plots that you subsequently create will fill the slots in the top row from left to right, followed by those in the second row and so on.

For example, to arrange four plots in a two-by-two grid, use the command:

```
> par(mfrow=c(2,2))
```

Then you can create up to four plots of any type to fill each of the slots, as shown in Figure 9-18:

```
> hist(iris$Sepal.Length)
> qqnorm(iris$Sepal.Length)
> pie(summary(iris$Species))
> plot(Petal.Length~Sepal.Length, iris)
```

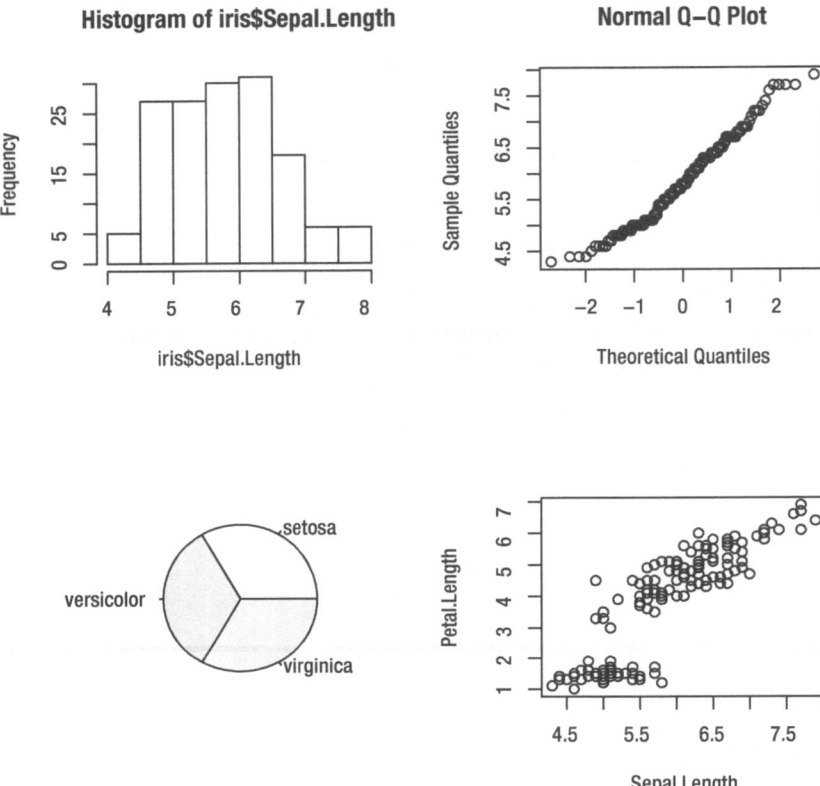

Figure 9-18. *Multiple plots*

When you create a fifth plot, a new image is started. The graphics device will continue to display multiple plots for the remainder of the session, or until you reverse it with the command:

```
> par(mfrow=c(1,1))
```

Changing the Default Plot Settings

So far, this chapter has explained how you can make modifications to a specific plot. However, you may want to change the settings so that they apply to all plots.

The par function allows you to change the default plotting style. For example, to change the default plotting symbol to be a triangle and the default title color to red, use the command:

```
> par(pch=2, col.main="red")
```

Any plots that you subsequently create will have red title text and use a triangle as the plotting symbol (if applicable). These changes are applied for the remainder of the session, or until they are overwritten.

R allows almost every aspect of the plots to be customized. Enter the command help(par) to see a full list of settings. Some arguments that you cannot change with the par function include main, sub, xlab, ylab, xlim, and ylim, which apply to individual plots only.

Summary

You should now be able to make your plots more informative by adding appropriate titles, labels, text, legends, and other items. You should also be able make them more visually appealing by adjusting aspects such as the colors and styles of lines and symbols. If necessary, you should be able to overlay several groups of data onto one plot, or display several plots in one image.

This table summarizes the most important arguments and commands covered in this chapter.

Task	Argument or command
Add title	main="*Title Text*"
Add subtitle	sub="*Subtitle text*"
Add axis labels	xlab="*X axis label*", ylab="*Y axis label*"
Change axis limits	xlim=c(*xmin*, *xmax*), ylim=c(*ymin*, *ymax*)
Change plotting color	col="red"
Change plotting symbol	pch=2
Change plotting symbol size	cex=2
Change line type	lty=2
Change line width	lwd=2
Change shading density	density=20
Add vertical line to plot	abline(v=2)
Add horizontal line to plot	abline(h=2)
Add straight line to plot	abline(a=*intercept*, b=*slope*)
Add line segment to plot	segments(*x1*,*y1*,*x2*,*y2*)
Add curve to plot	curve(*x^3*, add=T)
Add text to plot	text(*x*, *y*, "*Text String*")
Add grid to plot	grid()
Add arrow to plot	arrows(*x1*, *y1*, *x2*, *y2*)
Add points to plot	points(*dataset$variable*)
Add lines to plot	lines(*dataset$variable*)
Add legend to plot	legend(*x*, *y*, legend=c("*label1*", "*label2*", "*label3*"), ...)
Display multiple plots	par(mfrow=c(*rows*, *columns*))
Change default plot settings	par(...)

In the next chapter, we will leave plotting behind and move on to look at hypothesis testing.

■ ■ ■

CHAPTER 10

Hypothesis Testing

Hypothesis testing is useful because it allow you to use your data as evidence to support or refute a claim, and helps you to make decisions. This chapter explains how to perform some of the most commonly used types of hypothesis test.

Most of the tests discussed in this chapter are used to compare sample means, but we will also look at tests for comparing sample variances. Both parametric and nonparametric tests are covered. Parametric tests make some assumptions about the distribution of the data, usually that the data follows a normal distribution. Nonparametric tests don't make this kind of assumption, so are suitable in a wider range of situations.

You will learn how to:

- use a t-test to compare the means of two samples, or to compare one sample with a constant value

- perform a Wilcoxon rank-sum test (the non-parametric alternative to the t-test)

- use an analysis of variance to compare the means of three or more samples

- perform a Kruskal-Wallis test (the nonparametric alternative to the analysis of variance)

- perform pairwise comparisons using multiple comparison methods such as Tukey's HSD, Bonferroni and pairwise Wilcoxon rank-sum test

- compare the variances of two or more samples using an F-test or Bartlett's test

There are also several hypothesis tests that are covered in other chapters. The hypothesis test for correlation and the Shapiro-Wilk and Kolmogorov-Smirnov tests for fit to a distribution are covered in Chapter 5. Tests for tabular data, such as the chi-square test and Fisher's test, can be found in Chapter 6.

This chapter uses the iris, PlantGrowth, and sleep datasets (included with R) and the bottles, brains, and grades1 datasets (available with the downloads for this book). To view more details about the dataset included with R, enter help(*datasetname*). For further details about all other datasets, see Appendix C.

HYPOTHESIS TESTS

A hypothesis test uses a sample to test hypotheses about the population from which the sample is drawn. This helps you make decisions or draw conclusions about the population. A hypothesis test has the following components:

Null hypothesis (denoted H_0) is a hypothesis about the population from which a sample or samples are drawn. It is usually a hypothesis about the value of an unknown parameter such as the population mean or variance, e.g. H_0: *The population mean is equal to five*. The null hypothesis is adopted unless proven false.

Alternative hypothesis (denoted H_1 or H_A) is the hypothesis that will be accepted if there is enough evidence to reject the null hypothesis. This is generally the inverse of the null hypothesis, e.g. H_1: *The population mean is not equal to five*.

Test statistic is a statistic calculated from the sample values, which has a known distribution under the null hypothesis. It varies depending on the type of test and has no direct interpretation.

p-value gives the probability of observing the test statistic or something more extreme, assuming that the null hypothesis were true. If this is very small, then it suggests that the null hypothesis is not feasible, giving evidence in support of the alternative hypothesis.

Significance level (denoted α) is the cut-off point at which the null hypothesis is rejected. The significance level should be determined before beginning the test. Usually a significance level of 0.05 is used, but 0.01 and 0.1 are also popular choices. If a significance level of 0.05 is used, then we reject the null hypothesis in favor of the alternative hypothesis only if the p-value is less than 0.05. Otherwise, no conclusion is drawn. Choosing a significance level of 0.05 means that if the null hypothesis were true, there would be a 5% chance of incorrectly rejecting it (i.e., making a type I error).

Student's T-Tests

The Student's t-test is used to test hypotheses about the mean value of a population or two populations. A t-test is suitable if the data is believed to be drawn from a normal distribution. If the samples are large enough (i.e., at least 30 values per sample), then the t-test can be used even if the data is not normally distributed. This is because even if a random variable is not normally distributed, the mean values of a series of samples drawn from the distribution will tend to be normally distributed around the population mean, provided the samples are sufficiently large (the central limit theorem). If your data does not satisfy either of these criteria, then use the Wilcoxon rank-sum test instead, covered later in this chapter.

There are three types of t-test:

> **One-sample t-test** is used to compare the mean value of a sample with a constant value denoted μ_0. It has the null hypothesis that the population mean is equal to μ_0, and the alternative hypothesis that it is not.

> **Two-sample t-test** is used to compare the mean values of two independent samples, to determine whether they are drawn from populations with equal means. It has the null hypothesis that the two means are equal, and the alternative hypothesis that they are not equal.

> **Paired t-test** is used to compare the mean values for two samples, where each value in one sample corresponds to a particular value in the other sample. It has the null hypothesis that the two means are equal, and the alternative hypothesis that they are not equal.

The t-test can also be performed with a one-sided alternative hypothesis, which is known as a *one-tailed test*. For a one-sample t-test, the one-sided alternative hypothesis is either that the mean is greater than μ_0 or that the mean is less than μ_0. For a two-sample or paired t-test, the one-sided alternative hypothesis is either that the mean of the first population is greater than the mean of the second population, or that the mean of the first population is less than the mean of the second population.

The following sections explain how to perform each type of t-test in R.

One-Sample T-Test

You can perform a one-sample t-test with the t.test function. To compare a sample mean with a constant value mu0, use the command:

```
> t.test(dataset$sample1, mu=mu0)
```

The mu argument gives the value with which you want to compare the sample mean. It is optional and has a default value of 0.

By default, R performs a two-tailed test. To perform a one-tailed test, set the alternative argument to "greater" or "less":

```
> t.test(dataset$sample1, mu=mu0, alternative="greater")
```

A 95% confidence interval for the population mean is included with the output. To adjust the size of the interval, use the conf.level argument:

```
> t.test(dataset$sample1, mu=mu0, conf.level=0.99)
```

EXAMPLE 10-1.
ONE-TAILED, ONE-SAMPLE T-TEST USING THE BOTTLES DATA

A bottle filling machine is set to fill bottles with soft drink to a volume of 500 milliliters. The actual volume is known to follow a normal distribution. The manufacturer believes the machine is under-filling bottles. A sample of 20 bottles is taken and the volume of liquid inside is measured. The results are given in the bottles dataset, which is available from the website.

To calculate the sample mean, use the command:

```
> mean(bottles$Volume)
```

```
[1] 491.5705
```

Suppose that you want to use a one-sample t-test to determine whether the bottles are being consistently under filled, or whether the low mean volume for the sample could be the result of random variation. A one-sided test is suitable because the manufacturer is specifically interested in knowing whether the volume is *less* than 500 milliliters. The test has the null hypothesis that the mean filling volume is equal to 500 milliliters, and the alternative hypothesis that the mean filling volume is less than 500 milliliters. A significance level of 0.01 is to be used.

To perform the test, use the command:

```
> t.test(bottles$Volume, mu=500, alternative="less", conf.level=0.99)
```

This gives the following output:

```
        One Sample t-test

data:  bottles$Volume
t = -1.5205, df = 19, p-value = 0.07243
alternative hypothesis: true mean is less than 500
99 percent confidence interval:
    -Inf 505.6495
sample estimates:
mean of x
 491.5705
```

From the output, we can see that the mean bottle volume for the sample is 491.6 ml. The one-sided 99% confidence interval tells us that mean filling volume is likely to be less than 505.6 ml. The p-value of 0.07243 tells us that if the mean filling volume of the machine were 500 ml, the probability of selecting a sample with a mean volume less than or equal to this one would be approximately 7%.

Because the p-value is not less than the significance level of 0.01, we cannot reject the null hypothesis that the mean filling volume is equal to 500 ml. This means that there is no evidence that the bottles are being underfilled.

Two-Sample T-Test

You can use the t.test function to perform a two-sample t-test using data in both stacked and unstacked forms. Your data is in stacked form if all the data values are stored in one variable, and a second variable gives the name or number of the sample to which each observation belongs. Your data is in unstacked form if the values for each sample are held in separate variables. If you are unsure which form your data is in, consider Figure 10-1, which shows an example of a dataset in both stacked and unstacked form.

	ClassA	ClassB	ClassC
1	87	83	79
2	64	97	86
3	89	96	81
4	82	99	80
5	59	92	84

(a)grades1 (unstacked form)

	Result	Class
1	87	ClassA
2	64	ClassA
3	89	ClassA
4	82	ClassA
5	59	ClassA
6	83	ClassB
7	97	ClassB
8	96	ClassB
9	99	ClassB
10	92	ClassB
11	79	ClassC
12	86	ClassC
13	81	ClassC
14	80	ClassC
15	84	ClassC

(b)grades2 (stacked form)

Figure 10-1. *The same dataset in stacked and unstacked forms*

To perform a two-sample t-test with data in stacked form, use the command:

```
> t.test(values~groups, dataset)
```

where values is the name of the variable containing the data values and groups is the variable containing the sample names. If the grouping variable has more than two levels, then you must specify which two groups you want to compare:

```
> t.test(values~groups, dataset, groups %in% c("Group1", "Group2"))
```

If your data is in unstacked form, use the command:

```
> t.test(dataset$sample1, dataset$sample2)
```

By default, R uses separate variance estimates when performing two-sample and paired t-tests. If you believe the variances for the two groups are equal, you can use the pooled variance estimate. An F-test or Bartlett's test (see "Hypothesis Tests for Variance" later in this chapter) can help to determine whether this is the case. To use the pooled variance estimate, set the var.equal argument to T:

```
> t.test(values~groups, dataset, var.equal=T)
```

To perform a one-tailed test, set the alternative argument to "greater" or "less":

```
> t.test(values~groups, dataset, alternative="greater")
```

When the alternative argument is set to "greater", the alternative hypothesis for the test is that the mean for the first group is greater than the mean for the second group. If you are using stacked data, you may need to use the levels function to check which are the first and second groups (see "Working with Factor Variables" in Chapter 3). Similarly, setting it to "less" gives an alternative hypothesis that the mean for the first group is less than the mean for the second group.

A 95% confidence interval for the difference in means is included with the output. You can adjust the size of this interval with the conf.level argument:

```
> t.test(values~groups, dataset, conf.level=0.99)
```

EXAMPLE 10-2.
TWO-SAMPLE T-TEST USING THE IRIS DATA

Suppose that you wish to use a two-sample t-test to determine whether there is any real difference in mean sepal width for the Versicolor and Virginica species of iris. You can assume that sepal width is normally distributed, and that the variance for the two groups is equal. The null hypothesis for the test is that there is no real difference in mean sepal width for the two species, and the alternative hypothesis is that there is a difference.

The data is in stacked form, so perform the test with the command:

```
> t.test(Sepal.Width~Species, iris, Species %in% c("versicolor", "virginica"), var.equal=T)
```

The output is shown here.

```
        Two Sample t-test

data:  Sepal.Width by Species
t = -3.2058, df = 98, p-value = 0.001819
alternative hypothesis: true difference in means is not equal to 0
95 percent confidence interval:
 -0.33028246 -0.07771754
sample estimates:
mean in group versicolor  mean in group virginica
                   2.770                    2.974
```

The 95% confidence interval for the difference is −0.33 to −0.08, meaning that the mean sepal width for the Versicolor species is estimated to be between 0.08 and 0.33 cm less than for the Virginica species.

The p-value of 0.001819 is less than the significance level of 0.05, so we can reject the null hypothesis that the mean sepal width is the same for the Versicolor and Virginica species in favor of the alternative hypothesis that the mean sepal width is different for the two species.

This example is continued in Example 10-9, where an F-test is used to check the assumption of equal variance (see "F-Test").

Paired T-Test

You can perform a paired t-test by setting the paired argument to T. If your data is in stacked form, use the command:

```
> t.test(values~groups, dataset, paired=T)
```

Your data must have the same numbers of observations in each group, so that there is a one-to-one correspondence between the samples. R matches the first value from the first sample with the first value from the second sample.

For data in unstacked form, use the command:

```
> t.test(dataset$sample1, dataset$sample2, paired=T)
```

As for the two-sample test, you can adjust the test using the alternative, conf.level and var.equal arguments. So to perform a test with the alternative hypothesis that the mean for the first group is less than the mean for the second group, and that uses a pooled variance estimate, use the command:

```
> t.test(values~groups, dataset, paired=T, alternative="less", var.equal=T)
```

EXAMPLE 10-3.
PAIRED T-TEST USING THE BRAINS DATA

Consider the `brains` dataset shown in Figure 10-2, which gives brain volumes (in cubic centimetres) of the first and second-born twins for ten sets of monozygotic twins.

	Pair	Twin1	Twin2
1	1	1005	963
2	2	1035	1027
3	3	1281	1272
4	4	1051	1079
5	5	1034	1070
6	6	1079	1173
7	7	1104	1067
8	8	1439	1347
9	9	1029	1100
10	10	1160	1204

Figure 10-2. *The brains dataset, giving brain volume data from the article "Brain Size, Head Size, and IQ in Monozygotic Twins," by Tramo et al. (see Appendix C for more details)*

Suppose that you wish to use a t-test to determine whether brain volume is related to birth order. Brain volume is assumed to follow a normal distribution. The data is naturally paired because the first twin from a birth corresponds to the second twin in the same birth, so a paired t-test is suitable. Because differences in either direction are of interest, a two-tailed test is used. The null hypothesis for the test is that there is no difference in mean brain volume for first- and second-born twins, and the alternative hypothesis is that the mean brain volume for first-born twins is different to that of second-born twins.

To perform the test, use the command:

```
> t.test(brains$Twin1, brains$Twin2, paired=T)
```

The output is shown here.

```
        Paired t-test

data:  brains$Twin1 and brains$Twin2
t = -0.4742, df = 9, p-value = 0.6466
alternative hypothesis: true difference in means is not equal to 0
95 percent confidence interval:
 -49.04566  32.04566
sample estimates:
mean of the differences
                -8.5
```

The mean difference in brain size is estimated at −8.5, meaning that the brain size of the first born twins was an average of 8.5 cubic centimeters less than that of their second born sibling. The confidence interval for the difference is −49 to 32 cubic centimeters.

The p-value of 0.6466 tells us that if the mean brain size for first- and second-born twins is the same, the probability of observing a difference equal or greater in magnitude to that in our sample is 65%. As the p-value is not less than the significance level of 0.05, the null hypothesis cannot be rejected. This means there is no evidence that brain size is related to birth order in twins.

Wilcoxon Rank-Sum Test

The Wilcoxon rank-sum test (also known as the Mann-Whitney U test) allows you to test hypotheses about one or two sample means. It is a nonparametric alternative to the Student's t-test, which is suitable even when the distribution of the data is unknown and the samples are small. In parallel with the t-test there are one-sample, two-sample and paired forms. The alternative hypothesis can be either two-sided or one-sided.

You can perform a Wilcoxon rank-sum test with the `wilcox.test` function. The commands are similar to those used to perform a t-test.

To perform a one-sample test, use the command:

```
> wilcox.test(dataset$sample1, mu=mu0)
```

To perform a two-sample test with data in stacked form, use the command:

```
> wilcox.test(values~groups, dataset)
```

If the grouping variable has more than two levels then you must specify which two you want to compare:

```
> wilcox.test(values~groups, dataset, groups %in% c("Group1", "Group2"))
```

If your data is in unstacked form (with the values for each sample held in separate variables), use the command:

```
> wilcox.test(dataset$sample1, dataset$sample2)
```

For a paired test, set the `paired` argument to T:

```
> wilcox.test(values~groups, dataset, paired=T)
```

To perform a one-tailed test, set the `alternative` argument to "greater" or "less":

```
> wilcox.test(values~groups, dataset, alternative="greater")
```

By default, there is no confidence interval included with the output. To calculate a confidence interval for the population mean (for one-sample tests) or difference between means (for paired and two-sample tests), set the `conf.int` argument to T. The default size for the confidence intervals is 95%, but you can adjust this with the `conf.level` argument:

```
> wilcox.test(values~groups, dataset, conf.int=T, conf.level=0.99)
```

EXAMPLE 10-4.
PAIRED WILCOXON RANK-SUM TEST USING THE SLEEP DATA

Consider the `sleep` dataset, which is included with R. The data gives the results of an experiment in which 10 patients each took two different treatments and recorded the amount of additional sleep (compared to usual) that they experienced while receiving each of the treatments. The `extra` variable gives the increase in sleep (in hours per night), the `group` variable gives the treatment number (1 or 2), and the `ID` variable gives the patient number (1 to 10).

Suppose that you want to determine whether there is any real difference in the mean increase in sleep offered by the two treatments. A Wilcoxon rank-sum test is suitable because the distribution of additional sleep time is unknown, and the samples are small. A paired test is used because the additional sleep experienced by patient number x while taking drug 1 corresponds to the additional sleep experienced by patient number x while taking drug 2. The null hypothesis for the test is that there is no difference in mean additional sleep for the two treatments. The alternative hypothesis is that the mean additional sleep is different for the two treatments. A significance level of 0.05 will be used.

As the data is in stacked form, you can perform the test with the command:

```
> wilcox.test(extra~group, sleep, paired=T)
```

This gives the results:

```
        Wilcoxon signed rank test with continuity correction

data:  extra by group
V = 0, p-value = 0.009091
alternative hypothesis: true location shift is not equal to 0

Warning messages:
1: In wilcox.test.default(x = c(0.7, -1.6, -0.2, -1.2, -0.1, 3.4, 3.7,  :
  cannot compute exact p-value with ties
2: In wilcox.test.default(x = c(0.7, -1.6, -0.2, -1.2, -0.1, 3.4, 3.7,  :
  cannot compute exact p-value with zeroes
```

The p value of 0.009091 tells us that if the effect of both drugs were the same, there would be less than 1% chance of observing a difference in mean sleep increase as large as the one seen in this data. Because this is less than our significance level of 0.05, we can reject the null hypothesis that the additional sleep is the same for the two treatments. This means that there is evidence of a difference in effectiveness between the two treatments.

R has given a warning that there are ties in the data. This means that some observations have exactly the same value for the variable of interest, because the measurements have only been taken to one decimal place. For this reason, R is not able to calculate an exact p-value, and the results should be interpreted with caution. A more in-depth explanation of the effect of tied data on the Wilcoxon rank-sum test can be found on page 134 of *Nonparametric Statistics for the Behavioural Sciences*, Second Edition, by S. Siegel and N.J. Castellan (McGraw-Hill 1988).

Analysis of Variance

An analysis of variance (or ANOVA) allows you to compare the means of three or more independent samples. It is suitable when the values are drawn from a normal distribution and when the variance is approximately the same in each group. You can check the assumption of equal variance with a Bartlett's test (see "Hypothesis Tests for Variance" later in this chapter). The null hypothesis for the test is that the mean for all groups is the same, and the alternative hypothesis is that the mean is different for at least one pair of groups.

More complex models such as the two-way analysis of variance or the analysis of covariance are covered in Chapter 11.

You can perform an analysis of variance with aov function. Because the ANOVA is a type of general linear model, you could also use the lm or glm functions as explained in Chapter 11. However, the aov function presents the results more conveniently.

The command takes the form:

```
> aov(values~groups, dataset)
```

where values is the name of the variable that holds the data values and groups is the variable that identifies which sample each observation belongs to. If your data is in unstacked form (with the values for each sample held in separate variables) then you will need to stack the data beforehand, as explained in Chapter 4 under "Stacking Data."

The results of the analysis consist of many components that R does not automatically display. If you save the results to an object as shown here, you can use further functions to extract the various elements of the output:

```
> aovobject<-aov(values~groups, dataset)
```

Once you have saved the results as an object, you can view the ANOVA table with the anova function:

```
> anova(aovobject)
```

To view the model coefficients, use the coef function:

```
> coef(aovobject)
```

To view confidence intervals for the coefficients, use the confint function:

```
> confint(aovobject)
```

EXAMPLE 10-5.
ANALYSIS OF VARIANCE USING THE PLANTGROWTH DATA

Consider the PlantGrowth dataset (included with R), which gives the dried weight of three groups of 10 batches of plants, where each group of 10 batches received a different treatment. The weight variable gives the weight of the batch and the groups variable gives the treatment received. Figure 10-3 shows a boxplot of the weight grouped by treatment group.

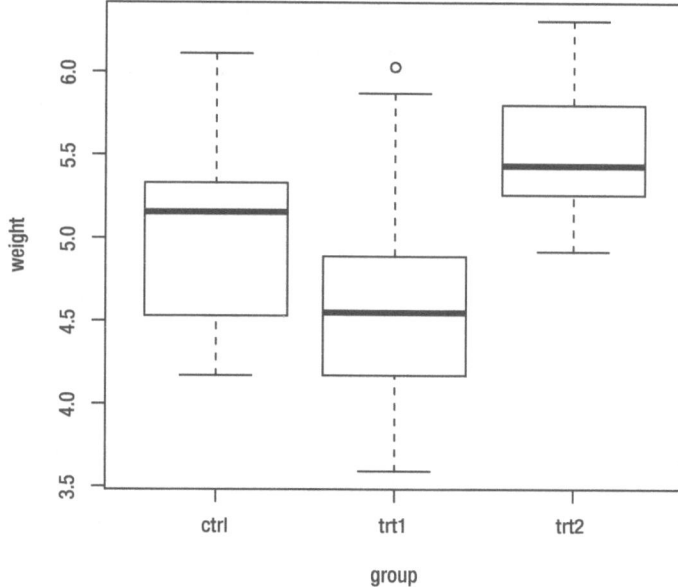

Figure 10-3. *Boxplot of weight grouped by treatment group for the* PlantGrowth *dataset*

You can see from the boxplot that there are some differences in plant growth (as measured by weight of batch) between the three groups. Suppose that you wish to perform an analysis of variance to determine whether these differences are statistically significant. You can assume that plant growth is normally distributed, and that the variance is the same for all three treatments. A significance level of 0.05 is used.

To perform the ANOVA and save the results to an object named plantanova, use the command:

```
> plantanova<-aov(weight~group, PlantGrowth)
```

To view the ANOVA table, use the command:

```
> anova(plantanova)
```

This gives the following output:

```
Analysis of Variance Table

Response: weight
          Df  Sum Sq Mean Sq F value  Pr(>F)
group      2  3.7663  1.8832  4.8461 0.01591 *
Residuals 27 10.4921  0.3886
---
Signif. codes:  0 '***' 0.001 '**' 0.01 '*' 0.05 '.' 0.1 ' ' 1
```

We can see that the p-value for the group effect is 0.01591. This means that if the effect of all three treatments were the same, we would have less than 2% chance of seeing differences between groups as large or larger than this. As the p-value is less than the significance level of 0.05, we can reject the null hypothesis that the mean growth is the same for all treatments, in favor of the alternative hypothesis that the mean growth is different for at least one pair of treatments.

To see the size of the treatment effects, use the command:

```
> coef(plantanova)
```

(Intercept)	grouptrt1	grouptrt2
5.032	-0.371	0.494

The output tells us that the control treatment gives an average weight of 5.032. The effect of treatment 1 (trt1) is to reduce weight by an average of −0.371 units compared to the control method, and the effect of treatment 2 (trt2) is to increase weight by an average of 0.494 units compared to the control method.

This example is continued in Example 10-7 (under "Tukey's HSD Test"), where pairwise t-tests are used to further investigate the treatment differences, and in Example 10-10, where a Bartlett's test is used to check the assumption of equal variance. Save the plantanova object, as it will be useful for these examples.

Kruskal-Wallis Test

The Kruskal-Wallis test allows you to compare the mean values of three or more samples. It is a nonparametric alternative to the analysis of variance, which can be used when the distribution of the values is unknown.

You can perform a Kruskal-Wallis test with the kruskal.test function. To perform the test with stacked data, use the command:

```
> kruskal.test(values~groups, dataset)
```

where the values variable contains the data values and the groups variable indicates to which sample each observation belongs.

For unstacked data (with samples in separate variables), nest the variables inside the list function:

```
> kruskal.test(list(dataset$sample1, dataset$sample2, dataset$sample3))
```

EXAMPLE 10-6.
KRUSKAL-WALLIS TEST USING GRADES1 DATA

Consider the `grades1` dataset, shown in Figure 10-4. It gives the test results (out of 100) of 15 students belonging to three different classes.

	ClassA	ClassB	ClassC
1	87	83	79
2	64	97	86
3	89	96	81
4	82	99	80
5	59	92	84

Figure 10-4. *grades1 dataset (see Appendix C for more details)*

Suppose that you want to use a Kruskal-Wallis test to determine whether there are any differences in the effectiveness of the teaching methods used by each of the three classes, as measured by the mean test results of the students. A significance level of 0.05 is used.

As the data is in unstacked form, you can perform the test with the command:

```
> kruskal.test(list(grades1$ClassA, grades1$ClassB, grades1$ClassC))
```

This gives the following output:

```
        Kruskal-Wallis rank sum test

data:  list(grades1$ClassA, grades1$ClassB, grades1$ClassC)
Kruskal-Wallis chi-squared = 6.6796, df = 2, p-value = 0.03544
```

From the output you can see that the p-value is 0.03544. As this is less than the significance level of 0.05, we can reject the null hypothesis that the mean score is equal for all classes. This means that there is evidence of a difference in effectiveness between the teaching methods used by the three classes.

This example is continued in Example 10-9, where pairwise Wilcoxon rank-sum tests are used to further investigate the differences between the three classes.

Multiple Comparison Methods

After performing an analysis of variance and finding that there are some differences between group means, you may want to perform a series of pairwise t-tests to identify differences between specific pairs of groups. However, when performing a large number of t-tests, the overall probability of incorrectly rejecting the null hypothesis for at least one of the tests (the type I error) is greater than the significance level used for the individual tests. Multiple comparison methods allow you to perform pairwise t-tests on three or more samples, while controlling the overall type I error.

There are a number of methods for performing multiple comparisons. Some of the most popular are the Tukey's honestly significant difference (HSD) test and the Bonferroni test.

This following sections describe how to perform several prominent multiple comparison methods in R.

Tukey's HSD Test

You can perform a Tukey's HSD test with the TukeyHSD function. If you have previously performed an analysis of variance using the aov, lm, or glm functions and saved the results to an object (as explained previously in "Analysis of Variance"), use the function directly:

```
> TukeyHSD(aovobject)
```

If you don't have an aov object and are using raw data, nest the aov function inside the TukeyHSD function:

```
> TukeyHSD(aov(values~groups, dataset))
```

The default overall confidence level is 0.95, but you can adjust it with the conf.level argument:

```
> TukeyHSD(aovobject, conf.level=0.99)
```

EXAMPLE 10-7.
TUKEY'S HSD TEST USING THE PLANTGROWTH DATA

In Example 10-5, an analysis of variance was used to help determine whether there are any differences in mean plant growth (measured by weight of batch) between the three treatment groups. The conclusion was that there is a difference in plant growth for at least one pair of treatments.

Suppose you wish to continue this analysis by using pairwise t-tests to determine which treatment groups have differences in plant growth. An overall significance level of 0.05 is used.

If you still have the plantanova aov object created in Example 10-5, you can perform the test with the command:

```
> TukeyHSD(plantanova)
```

If you no longer have the model object, you can perform the test from the raw data:

```
> TukeyHSD(aov(weight~group, PlantGrowth))
```

The output is shown here:

```
Tukey multiple comparisons of means
  95% family-wise confidence level

Fit: aov(formula = weight ~ group, data = PlantGrowth)

$group
            diff        lwr       upr       p adj
trt1-ctrl -0.371 -1.0622161 0.3202161 0.3908711
trt2-ctrl  0.494 -0.1972161 1.1852161 0.1979960
trt2-trt1  0.865  0.1737839 1.5562161 0.0120064
```

The diff column shows the difference in sample means for the two groups, and the lwr and upr columns give a 95% confidence interval for the difference. The p adj column gives the p-value for a t-test between the two groups, adjusting for multiple t-tests.

Comparing the p-values to our significance level of 0.05, we can see that the comparison trt2-trt1 is statistically significant. This means that the plant growth for treatment 1 is significantly different from the growth for treatment 2. Treatment 1 and treatment 2 are not significantly different from the control.

Treatment 2 is estimated to give 0.865 units more growth than treatment 1. The 95% confidence interval for the difference in growth is 0.174 to 1.556.

Other Pairwise T-Tests

The pairwise.t.test function allows you to perform pairwise t-tests using a number of other multiple comparison methods. To perform pairwise t-tests using the Bonferroni adjustment, use the command:

```
> pairwise.t.test(dataset$values, dataset$groups, p.adj="bonferroni")
```

Other possible options for the p.adj argument include holm (the default method), hochberg, and hommel. Enter help(p.adjust) to view a complete list.

EXAMPLE 10-8.
PAIRWISE COMPARISONS WITH BONFERRONI ADJUSTMENT
USING THE PLANTGROWTH DATA

Continuing the PlantGrowth example, suppose that you wish to try performing pairwise t-tests using the Bonferroni adjustment for comparison:

```
> pairwise.t.test(PlantGrowth$weight, PlantGrowth$group, p.adj="bonferroni")
```

The results are shown here:

```
        Pairwise comparisons using t tests with pooled SD

data:  PlantGrowth$weight and PlantGrowth$group

     ctrl  trt1
trt1 0.583 -
trt2 0.263 0.013

P value adjustment method: bonferroni
```

The output shows p-values for each of the comparisons. From the output, we can see that the comparison of treatment 1 (trt1) and treatment 2 (trt2) has a p-value of 0.013, which is statistically significant at the 0.05 level. The comparison between the control and treatment 1, and between the control and treatment 2 were not statistically significant. This is consistent with the results of the Tukey's HSD test in the previous example.

Pairwise Wilcoxon Rank-Sum Tests

There is also a function called `pairwise.wilcox.test`, which allows you to perform pairwise Wilcoxon rank-sum tests. This is useful for identifying differences between individual pairs of groups after performing a Kruskal-Wallis test (described earlier in this chapter).

To perform the test, use the command:

```
> pairwise.wilcox.test(dataset$values, dataset$groups)
```

As when performing pairwise t-tests, you can adjust the multiple comparison method with the `p.adj` argument.

Hypothesis Tests for Variance

A hypothesis test for variance allows you to compare the variance of two or more samples to determine whether they are drawn from populations with equal variance. The tests have the null hypothesis that the variances are equal and the alternative hypothesis that they are not. These tests are useful for checking the assumptions of a t-test or analysis of variance.

Two types of test for variance are covered in this section:

F-test allows you to compare the variance of two samples. It is suitable for normally distributed data.

Bartlett's test allows you to compare the variance of two or more samples. It is suitable for normally distributed data.

The following sections explain how to perform both the F-test and Bartlett's tests in R.

F-Test

You can perform an F-test with the `var.test` function. If your data is in stacked form, use the command:

```
> var.test(values~groups, dataset)
```

If the groups variable has more than two levels, then you must specify which two you want to compare:

```
> var.test(values~groups, dataset, groups %in% c("Group1", "Group2"))
```

For data in unstacked form (with the samples in separate variables), use the command:

```
> var.test(dataset$sample1, dataset$sample2)
```

EXAMPLE 10-10.
F-TEST USING THE IRIS DATASET

In Example 10-2, we used a t-test to compare the mean sepal width for the Versicolor and Virginica species of iris, using a pooled variance estimate. One of the assumptions made for this test was that the variance in sepal width is the same for both species.

Suppose that you want to use an F-test to help determine whether the variance in sepal width is the same for the two species. A significance level of 0.05 will be used.

To perform the test, use the command:

```
> var.test(Sepal.Width~Species, iris, Species %in% c("versicolor", "virginica"))
```

The output is shown here.

```
        F test to compare two variances

data:  Sepal.Width by Species
F = 0.9468, num df = 49, denom df = 49, p-value = 0.849
alternative hypothesis: true ratio of variances is not equal to 1
95 percent confidence interval:
 0.5372773 1.6684117
sample estimates:
ratio of variances
          0.9467839
```

The p-value of 0.849 is not less than the significance level of 0.05, so we cannot reject the null hypothesis that the variance for the two groups is equal. There is no evidence to suggest that the variance in sepal width is different for the Versicolor and Virginica species.

Bartlett's Test

You can perform the Bartlett's test with the `bartlett.test` function. If your data is in stacked form, use the command:

```
> bartlett.test(values~groups, dataset)
```

Unlike the F-test, the Bartlett's test allows you to compare the variance of more than two groups. However, if required, you can still select a subset of groups to compare:

```
> bartlett.test(values~groups, dataset, groups %in% c("Group1", "Group2"))
```

If you have data in unstacked form (with the samples in separate variables), nest the variables inside the `list` function:

```
> bartlett.test(list(dataset$sample1, dataset$sample2, dataset$sample3))
```

EXAMPLE 10-11.
BARTLETT'S TEST USING THE PLANTGROWTH DATA

In Example 10-5, we used an analysis of variance to compare the mean weight of plant batches for the three treatment groups. One of the assumptions made for this test was that the variance in weight is the same for all treatment groups.

Suppose that you want to use Bartlett's test to determine whether the variance in weight is the same for all treatment groups. A significance level of 0.05 will be used.

To perform the test, use the command:

```
> bartlett.test(weight~group, PlantGrowth)
```

This gives the output:

```
        Bartlett test of homogeneity of variances

data:  weight by group
Bartlett's K-squared = 2.8786, df = 2, p-value = 0.2371
```

From the output, we can see that the p-value of 0.2371 is not less than the significance level of 0.05. This means that we cannot reject the null hypothesis that the variance is the same for all treatment groups. This means that there is no evidence to suggest that the variance in plant growth is different for the three treatment groups.

Summary

You should now be able to compare the means of two samples using the t-test or Wilcoxon rank-sum test, and use an analysis of variance or Kruskal-Wallis test to compare the means of three or more samples. You should be able to perform pairwise comparisons of three or more groups using an appropriate method. Finally, you should be able to use an F-test or Bartlett's test to compare the variances of two groups.

This table summarizes the most important commands covered.

Test Type	Command
One-sample t-test	t.test(*dataset$sample1*, mu=*mu0*)
Two-sample t-test	t.test(*values~groups, dataset*)
	t.test(*dataset$sample1, dataset$sample2*)
Paired t-test	t.test(*values~groups, dataset*, paired=T)
	t.test(*dataset$sample1, dataset$sample2*, paired=T)
One-sample Wilcoxon rank-sum test	wilcox.test(*dataset$sample1*, mu=*mu0*)
Two-sample Wilcoxon rank-sum test	wilcox.test(*values~groups, dataset*)
	wilcox.test(*dataset$sample1, dataset$sample2*)

(*continued*)

Test Type	Command
Paired Wilcoxon rank-sum test	wilcox.test(*values~groups*, *dataset*, paired=T)
	wilcox.test(*dataset$sample1*, *dataset$sample2*, paired=T)
Analysis of variance	aov(*values~groups*, *dataset*)
Kruskal-Wallis test	kruskal.test(*values~groups*, *dataset*)
	kruskal.test(list(*dataset$sample1*, *dataset$sample2*, *dataset$sample3*))
Tukey's HSD test	tukeyHSD(*aovobject*)
	tukeyHSD(aov(*values~groups*, *dataset*))
Pairwise t-tests	paired.t.test(*dataset$values*, *dataset$groups*, method="bonferroni")
Pairwise Wilcoxon rank-sum tests	paired.wilcox.test(*dataset$values*, *dataset$groups*)
F-test	var.test(*values~groups*, *dataset*)
	var.test(*dataset$sample1*, *dataset$sample2*)
Bartlett's test	bartlett.test(*values~groups*, *dataset*)
	bartlett.test(list(*dataset$sample1*, *dataset$sample2*))

In the next chapter, we will look at how to build regression models and other types of general linear model.

CHAPTER 11

Regression and General Linear Models

Model building helps you to understand the relationships between variables and to make predictions about future observations. This chapter explains how to build regression models and other models in the general linear model family.

You will learn how to:

- build simple linear regression, multiple linear regression and polynomial regression models
- include interaction terms, transformed variables, and factor variables in a model
- add or remove terms from an existing model
- perform the stepwise, forward, and backward model selection procedures
- assess how well a model fits the data
- interpret model coefficients
- represent a model graphically with a line or curve of best fit
- check the validity of a model using diagnostics such as the residuals, deviance, and Cook's distances
- use your model to make predictions about new data

This chapter uses the trees dataset (included with R) and the powerplant, concrete and people2 datasets (which are available with the downloads for this book and described in Appendix C).

GENERAL LINEAR MODELS

A general linear model is used to predict the value of a continuous variable (known as the *response* variable) from one or more explanatory variables. A general linear model takes the form:

$y = \beta_0 + \beta_1 x_1 + \beta_2 x_2 + \dots \dots + \beta_n x_n + \varepsilon$

where y is the response variable, x_i are the explanatory variables, β_i are coefficients to be estimated and ε represents the random error.

The explanatory variables can be either continuous or categorical, and they can include cross products, polynomials and transformations of other variables.

The random errors are assumed to be independent, to follow a normal distribution with a mean of zero, and to have the same variance for all values of the explanatory variables.

Simple linear regression, multiple linear regression, polynomial regression, analysis of variance, two-way analysis of variance, analysis of covariance, and experimental design models are all types of general linear model.

Building the Model

You can build a regression model or general linear model with either the lm function or the glm function. When used for this purpose, these functions do the same thing and give similar output. In this book we will use the lm function for building general linear models, but be aware that you may also see the glm function in use.

Simple Linear Regression

To build a simple linear regression model with an explanatory variable named var1 and a response variable named resp, use the command:

```
> lm(resp~var1, dataset)
```

You don't need to specify an intercept term (or constant term) in your model because R includes one automatically. To build a model without an intercept term, use the command:

```
> lm(resp~-1+var1, dataset)
```

When you build a model with the lm function, R displays the coefficient estimates. However, there are more components to the output that are not displayed, such as summary statistics, residuals, and fitted values. You can save the all of the output to an object, as shown here. Later in the chapter, you will learn how to access the various components of the model output.

```
> modelname<-lm(resp~var1, dataset)
```

EXAMPLE 11-1. SIMPLE LINEAR REGRESSION USING THE TREES DATA

In the "Pearson's Correlation Coefficient" section in Chapter 5, we saw that there is a correlation between tree girth and tree volume. Suppose that you want to build a simple linear regression model to predict a tree's volume from its girth. To build the model, use the command:

```
> lm(Volume~Girth, trees)
```

```
Call:
lm(formula = Volume ~ Girth, data = trees)

Coefficients:
(Intercept)        Girth
    -36.943        5.066
```

From the output, you can see that the model formula is:

Volume=-36.943+5.066×Girth

This means that a tree with a girth of 10 inches has an expected volume of −36.943+5.066×10=13.717 cubic feet. For every additional inch of girth, the expected volume increases by 5.066 cubic feet.

To save the model output as an object, use the command:

```
> treemodel<-lm(Volume~Girth, trees)
```

The treemodel object will be used later in the chapter.

Multiple Linear Regression

To include several explanatory variables in a model, separate them with the plus sign:

```
> modelname<-lm(resp~var1+var2+var3, dataset)
```

EXAMPLE 11-2.
MULTIPLE LINEAR REGRESSION USING THE POWERPLANT DATA

Consider the powerplant dataset, which is described in Appendix C and is available with the downloads for this book. The dataset has three variables. The Output variable gives the output of a gas electrical turbine in megawatts. The Pressure and Temp variables give temperature and pressure measurements inside the turbine.

To build a multiple linear regression model that predicts the output from the pressure and temperature, use the command:

```
> lm(Output~Pressure+Temp, powerplant)
```

```
Call:
lm(formula = Output ~ Pressure + Temp, data = powerplant)

Coefficients:
(Intercept)      Pressure          Temp
   -32.8620        0.1858       -0.9916
```

From the output you can see that the model is:

Output = -32.8620 + 0.1858×Pressure - 0.9916×Temperature

This tells us that for an increase in pressure of one millibar, the expected output of the turbine increases by 0.1858 megawatts. For an increase in temperature of one centigrade, the expected output decreases by 0.9916 megawatts.

Interaction Terms

To add an interaction term to a model, use a colon (:). For example, var1:var2 denotes the interaction between the two variables var1 and var2. This command builds a model with terms for two variables and their interaction:

```
> modelname<-lm(resp~var1+var2+var1:var2, dataset)
```

You can also use a colon to express third-order and higher interactions:

```
> modelname<-lm(resp~var1+var2+var3+var1:var2+var1:var3+var2:var3+var1:var2:var3, dataset)
```

As you can see, this notation becomes lengthy for models with many variables. R has two useful shorthand notations for interaction terms, which are the asterisk (*) notation and the hat (^) notation.

Use the asterisk notation to include a group of variables and all their possible interactions. For example, the command:

```
> modelname<-lm(resp~var1*var2*var3, dataset)
```

builds a model with terms for the three variable main effects (var1, var2, var3), the three second-order interactions (var1:var2, var1:var3, var2:var3), and the third-order interaction (var1:var2:var3). This is equivalent to the previous very lengthy formula.

The hat notation includes a set of variables as well as all the possible interactions up to a given order. For example, to include all main effects and second-order interactions (but not the third-order interaction), use the command:

```
> modelname<-lm(resp~(var1+var2+var3)^2, dataset)
```

This is equivalent to this command:

```
> modelname<-lm(resp~var1+var2+var3+var1:var2+var1:var3+var1:var2, dataset)
```

EXAMPLE 11-3. FACTORIAL EXPERIMENT USING THE CONCRETE DATA

Consider the concrete dataset, which is shown in Figure 11-1 and available from the website. It gives the results of an experiment to determine the effect of cement type (I or II), additive type (A or B), and additive dose (0.3%, 0.4% or 0.5%) on the density of concrete.

	Cement	Additive	Additive.Dose	Density
1	I	A	0.003	2.43
2	II	A	0.003	2.431
3	I	B	0.003	2.43
4	II	B	0.003	2.418
5	I	A	0.004	2.419
6	II	A	0.004	2.436
7	I	B	0.004	2.435
8	II	B	0.004	2.414
9	I	A	0.005	2.419
10	II	A	0.005	2.425
11	I	B	0.005	2.422
12	II	B	0.005	2.41

Figure 11-1. *The concrete dataset (see Appendix C for more details)*

To build a model to predict density that includes terms for all of the explanatory variables (cement type, additive type, and additive dose) and their interactions (second and third order), use the command:

```
> concmodel<-lm(Density~Cement*Additive*Additive.Dose, concrete)
```

The concmodel object will be used later in this chapter.

Polynomial Terms

To build a polynomial regression model with terms for var1, var1^2, and var1^3, use the command:

```
> modelname<-lm(resp~var1+I(var1^2)+I(var1^3), dataset)
```

Notice that the terms var1^2 and var1^3 are nested inside an I(). This is because the symbols ^, * and + have special meanings when used in a model formula, which can be confused with their usual arithmetic meanings of power, multiplication and addition. If you want to use these symbols for their usual arithmetic meanings, you must nest them inside the I function.

EXAMPLE 11-4.
POLYNOMIAL REGRESSION USING THE TREES DATASET

To build a model to predict tree volume that has terms for girth and girth^2, use the command:

```
> lm(Volume~Girth+I(Girth^2), trees)
```

```
Call:
lm(formula = Volume ~ Girth + I(Girth^2), data = trees)

Coefficients:
(Intercept)        Girth   I(Girth^2)
    10.7863      -2.0921       0.2545
```

From the result, you can see that the model formula is:

Volume = 10.7863 - 2.0921×Girth + 0.2545×Girth^2

To save the model as an object, use the command:

```
> polytrees<-lm(Volume~Girth+I(Girth^2), trees)
```

The polytrees object will be used later in this chapter.

Transformations

Simple variable transformations such as the log or square root transformations can be applied to the response or explanatory variables directly in the formula using the relevant function:

```
> modelname<-lm(log(resp)~var1, dataset)
```

For transformations that use the asterisk, hat, or plus symbols, nest them inside the I function:

```
> modelname<-lm(I(resp^2)~var1, dataset)
```

The Intercept Term

There may be occasions when you want to build a model without an intercept term. To do this, add -1 as a term in the model, as shown here. This tells R not to include an intercept term, which would otherwise be included automatically:

```
> modelname<-lm(resp~-1+var1+var2+var3, dataset)
```

You can also create a model that contains only the intercept term, known as the *null model*:

```
> modelname<-lm(resp~1, dataset)
```

Including Factor Variables

You can include categorical variables in your model in the same way as continuous variables. Before including a categorical variable, use the class function to check that it has the factor variable class, as explained in Chapter 3 under "Working with Factor Variables."

When you include a factor variable in a model, R treats the first level of the factor as the reference level. So for a factor with n level, the model will have n-1 coefficients that express the effect of the remaining n-1 levels relative to the reference level.

To check which is the first level for a factor variable, use the levels function:

```
> levels(dataset$variable)
```

You can change the reference level for a factor variable with the relevel function:

```
> dataset$variable<-relevel(dataset$variable, "reflevel")
```

It is possible to change the way that R uses the coefficients to express the effect of the factor by changing the *contrasts* for the factor variable. Every factor variable has a set of contrasts associated with it, which you can view and change with the contrasts function. Enter help(contrasts) for more details on how to change the contrasts for a factor variable, and enter help(contr.treatment) for a list of contrast options.

EXAMPLE 11-5.
MODEL WITH FACTOR VARIABLE, USING THE PEOPLE2 DATASET

Suppose that you want to use the people2 dataset to build a model to predict a person's height from their hand span and eye color.

You can build the model with the command:

```
> lm(Height~Hand.Span+Eye.Color, people2)
```

```
Call:
lm(formula = Height ~ Hand.Span + Eye.Color, data = people2)

Coefficients:
    (Intercept)        Hand.Span   Eye.ColorBrown   Eye.ColorGreen
        82.8902           0.4456          -3.6233          -4.1924
```

From the output, we can see that formula has two coefficients that express the effect of brown and green eyes on height relative to blue eyes. So for people with blue eyes, the model formula is:

Height = 82.8902+0.4456×Hand Span

The formula for people with brown eyes is:

Height = 82.8902 + 0.4456×Hand Span -3.6233

The formula for people with green eyes is:

Height = 82.8902 + 0.4456×Hand Span -4.1924

Updating a Model

Once you have built a model, you may want to add or remove a term from the model to see how the new model compares with the previous one. The update function allows you build a new model by adding or removing terms from an existing model. This is useful when working with models that have many terms, as it means that you don't have to retype the entire model specification every time you add or remove a term.

Suppose that you have built a model and saved it to an object named model1:

```
> model1<-lm(resp~var1+var2+var3+var4, dataset)
```

To create a new model named model2 by adding an additional term to model1 (such as the interaction var1:var2), use the command:

```
> model2<-update(model1, ~.+var1:var2)
```

Similarly, you can remove a term from the model:

```
> model2<-update(model1, ~.-var4)
```

To check that the new model has the formula you expect, use the formula function:

```
> formula(model2)
```

EXAMPLE 11-6. UPDATING THE CONCMODEL MODEL

In Example 11-3, we built the concmodel model with the command:

```
> concmodel<-lm(Density~Cement*Additive*Additive.Dose, concrete)
```

To create a new model by removing the three-way interaction from the concmodel model, use the command:

```
> concmodel2<-update(concmodel, ~.-Cement:Additive:Additive.Dose)
```

Check the model formula for the new model with the command:

```
> formula(concmodel2)
```

```
Density ~ Cement + Additive + Additive.Dose + Cement:Additive +
    Cement:Additive.Dose + Additive:Additive.Dose
```

You can see that the three-way interaction (`Cement:Additive:Additive.Dose`) has now been removed from the model.

Stepwise Model Selection Procedures

Stepwise model selection procedures are algorithms designed to simplify the process of finding a model that explains a large amount of variation while including as few terms as possible. They are useful when dealing which large models with many potential terms. Popular stepwise selection procedures include forward selection, backward elimination, and general stepwise selection.

The `step` function allows you to perform forward, backward, and stepwise model selection in R. The function takes an `lm` or `glm` model object as input, which should be the full model from which you want to select a subset of terms.

Suppose that you have created a large model such as the one shown here, which includes a total of fifteen terms: four main effects, six second-order interactions, four third-order interactions, and one fourth-order interaction:

> *model1<-lm(resp~var1*var2*var3*var4, dataset)*

Once you have created the model, perform the stepwise selection procedure with the command:

> *model2<-step(model1)*

By default, the `step` function uses the general stepwise selection method. To use the backward or forward methods, set the `direction` argument to `"backward"` or `"forward"`:

> *model2<-step(model1, direction="backward")*

The newly created model object can be used in just the same way as any other model object. To see which terms have been kept in the new model, use the `formula` function:

> formula(*model2*)

EXAMPLE 11-7.
STEPWISE SELECTION USING THE CONCRETE DATA

In Example 11.3, we built the `concmod` model with the command:

> concmodel<-lm(Density~Cement*Additive*Additive.Dose, concrete)

To perform the stepwise selection procedure on this model, use the command:

> concmodel3<-step(concmodel)

To view the formula of the resulting model, use the command:

```
> formula(concmodel3)
```

```
Density ~ Cement + Additive + Additive.Dose + Cement:Additive
```

From the output, you can see that the stepwise selection procedure has removed the three-way interaction and two of the two-way interaction terms from the model, leaving the main effects of cement type, additive type, and additive dose, and the interaction between cement type and additive type.

Assessing the Fit of the Model

Before interpreting and using your model, you will need to determine whether it is a good fit to the data and includes a good combination of explanatory variables. You may also be considering several alternative models for your data and want to compare them.

MODEL FIT

The fit of a model is commonly measured in a few different ways. These include:

Coefficient of determination (R^2) gives an indication of how well the model is likely to predict future observations. It measures the portion of the total variation in the data that the model is able to explain. It takes values between 0 and 1. A value close to 1 suggests that the model will give good predictions, while a value close to 0 suggests that the model will make poor predictions.

Adjusted R-squared is similar to R^2, but makes an adjustment for the number of terms in the model.

Significance test for model coefficients tells you whether individual coefficient estimates are significantly different from 0, and hence whether the coefficients are contributing to the model. Consider removing coefficients with p-values greater than 0.05.

F-test tells you whether the model is significantly better at predicting compared with using the overall mean value as a prediction. For good models, the p-value will be less than 0.05. An F-test can also be used to compare two models. In this case, a p-value less than 0.05 tells you that the more complex model is significantly better than the simpler model.

To view some summary statistics for a model, use the summary function:

```
> summary(lmobject)
```

The summary function displays:

- the model formula
- quantiles for the residuals
- coefficient estimates with the standard error and a significance test for each
- the residual standard error and degrees of freedom
- the R^2 (multiple and adjusted)
- an F-test for model fit

Note that if you built your model with the glm function instead of the lm function, the output will be slightly different and will use the generalized linear model terminology.

To view an ANOVA table for the model, which shows the calculations behind the F-test, use the anova function:

```
> anova(lmobject)
```

You can also use the anova function to perform an F-test to compare a more complex model to a simpler model (the order of the models is not important):

```
> anova(model1, model2)
```

This only works for nested models. That is, all of the terms in the simpler model must also be found in the more complex model.

EXAMPLE 11-8.
MODEL SUMMARY STATISTICS FOR THE TREEMODEL MODEL

In Example 11-1, we created a simple linear regression model to predict tree volume from tree girth and saved the output to an object named treemodel.

To view summary statistics for the model, use the command:

```
> summary(treemodel)
```

This gives the following output:

```
Call:
lm(formula = Volume ~ Girth, data = trees)

Residuals:
    Min     1Q Median     3Q    Max
 -8.065 -3.107  0.152  3.495  9.587

Coefficients:
            Estimate Std. Error t value Pr(>|t|)
(Intercept) -36.9435     3.3651  -10.98 7.62e-12 ***
Girth         5.0659     0.2474   20.48  < 2e-16 ***
---
Signif. codes:  0 '***' 0.001 '**' 0.01 '*' 0.05 '.' 0.1 ' ' 1

Residual standard error: 4.252 on 29 degrees of freedom
Multiple R-squared:  0.9353,    Adjusted R-squared:  0.9331
F-statistic: 419.4 on 1 and 29 DF,  p-value: < 2.2e-16
```

The R^2 value of 0.9353 tells us that the model explains approximately 94% of the variation in tree volume. This suggests the model would be very good at predicting tree volume.

The hypothesis tests for the model coefficients tell us that the intercept and girth coefficients are significantly different from 0.

The p-value for the F-test is less than 0.05, which tells us that the model explains a significant amount of the variation in tree volume.

To view the ANOVA table, use the command:

```
> anova(treemodel)
```

```
Analysis of Variance Table

Response: Volume
          Df Sum Sq Mean Sq F value    Pr(>F)
Girth      1 7581.8  7581.8  419.36 < 2.2e-16 ***
Residuals 29  524.3    18.1
---
Signif. codes:  0 '***' 0.001 '**' 0.01 '*' 0.05 '.' 0.1 ' ' 1
```

The output shows the calculations behind the F-test.

EXAMPLE 11-9.
COMPARING THE TREEMODEL AND POLYTREES MODELS

Suppose that you want to compare treemodel (the simple linear regression model created in Example 11-1) and polytrees (the polynomial regression model created in Example 11-4) to determine whether the additional polynomial term in the polytrees model significantly improves the fit of the model.

To perform an F-test to compare the two models, enter the command:

```
> anova(treemodel, polytrees)
```

```
Analysis of Variance Table

Model 1: Volume ~ Girth
Model 2: Volume ~ Girth + I(Girth^2)
  Res.Df    RSS Df Sum of Sq      F    Pr(>F)
1     29 524.30
2     28 311.38  1    212.92 19.146 0.0001524 ***
---
Signif. codes:  0 '***' 0.001 '**' 0.01 '*' 0.05 '.' 0.1 ' ' 1
```

From the results, you can see that the p-value of 0.0001524 is less than 0.05. This means that the polytrees model fits the data significantly better than the treemodel model.

Coefficient Estimates

Coefficient estimates (or parameter estimates) are provided with the summary output, but you can also view them with the coef function, or the identical coefficients function:

```
> coef(lmobject)
```

view confidence intervals for the coefficient estimates, use the confint function:

```
> confint(lmobject)
```

The function gives 95% intervals by default, but you can adjust this with the level argument:

```
> confint(lmobject, level=0.99)
```

EXAMPLE 11-10.
COEFFICIENTS ESTIMATES FOR THE TREEMODEL MODEL

To view coefficient estimates for the treemodel, use the command:

```
> coef(treemodel)
```

```
(Intercept)        Girth
 -36.943459     5.065856
```

To view confidence intervals for the coefficient estimates, use the command:

```
> confint(treemodel)
```

```
                2.5 %      97.5 %
(Intercept) -43.825953 -30.060965
Girth         4.559914   5.571799
```

From the output, you can see that the 95% confidence interval for the intercept is −43.83 to −30.06, and the 95% confidence interval for the Girth coefficient is 4.56 to 5.57.

Plotting the Line of Best Fit

For simple linear regression models, you can create a scatter plot with a line of best fit superimposed over the data to help visualize the model. Use the plot and abline functions:

```
> plot(resp~var1, dataset)
> abline(coef(lmobject))
```

See the "Scatter Plots" section in Chapter 8 for more details about creating scatter plots with the plot function, and the "Adding straight Lines" section in Chapter 9 for more about adding straight lines with the abline function.

For polynomial regression models, you can superimpose a polynomial curve over the data. Use the curve function (see "Plotting a Function" in Chapter 8).

EXAMPLE 11-11.
SCATTER PLOT WITH LINE OF BEST FIT FOR THE TREEMODEL MODEL

To create a scatter plot with line of best fit for the `treemodel`, use the commands:

```
> plot(Volume~Girth, trees)
> abline(coef(treemodel))
```

The result is shown in Figure 11-2.

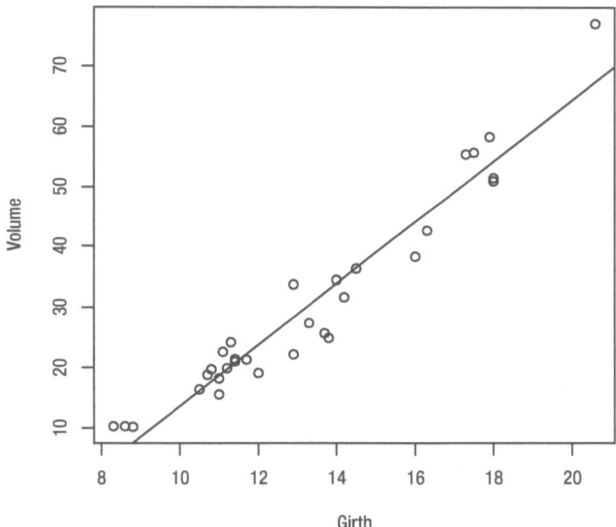

Figure 11-2. *Scatter plots with simple linear regression model superimposed*

EXAMPLE 11-12.
SCATTER PLOT WITH POLYNOMIAL CURVE FOR THE POLYTREES MODEL

In Example 11-4, we created a polynomial regression model named `polytrees` to predict tree volume from tree girth. To view the model coefficients for the `polytrees` model, use the `coef` function:

```
> coef(polytrees)
```

```
(Intercept)      Girth  I(Girth^2)
 10.7862655  -2.0921396   0.2545376
```

To create a scatter plot with curve superimposed, use the `curve` function:

```
> plot(Volume~Girth, trees)
> curve(10.7863-2.0921*x+0.2545*x^2, add=T)
```

Figure 11-3 shows the result.

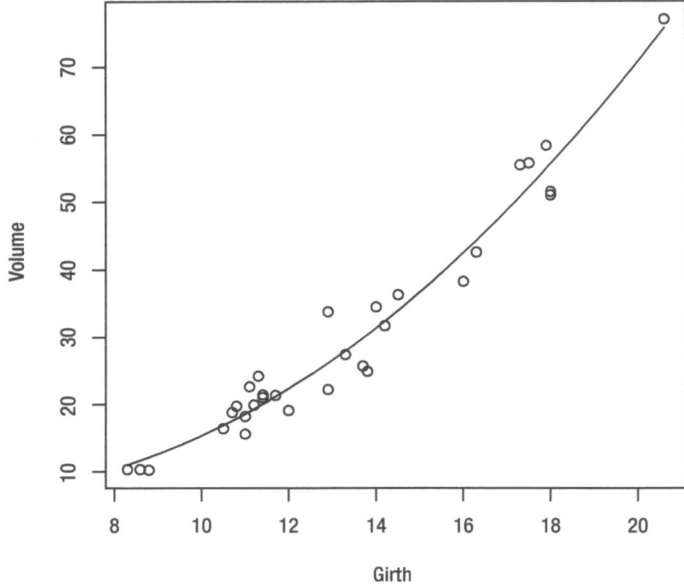

Figure 11-3. *Scatter plots with polynomial regression model superimposed*

Model Diagnostics

This section discusses the methods used to check the suitability of the model for the data and the reliability of the coefficient estimates. These include examining the model residuals and also measures of influence such as the leverage and Cook's distances for each of the observations.

Residual Analysis

The residuals for a given model are the set of differences between the observed values of the response variable, and the values predicted by the model (the fitted values). Standardized residuals and studentized residuals are types of residuals that have been adjusted to have a variance of one.

Examining the set of residuals for a model helps you to determine whether the model is appropriate. If the assumptions of a general linear model are met, then the residuals will be normally distributed and have constant variance. They will also be independent and will not follow any observable patterns. Residuals also help you to identify any observations that heavily influence the model.

To calculate raw residuals for a model, use the `residuals` function. There is also an identical function called `resid`. To calculate standardized residuals, use the `rstandard` function, and for studentized residuals, use the `rstudent` function:

```
> residuals(lmobject)
```

You can save the residuals to a new variable in your dataset, as shown here. This is useful if you want to plot the residuals against the response or explanatory variables:

```
> dataset$resids<-rstudent(lmobject)
```

Similarly, you can create a new variable containing fitted values with the fitted function:

```
> dataset$fittedvals<-fitted(lmobject)
```

Plotting the residuals is the easiest way to check for violations of the model assumptions. A normal probability plot or histogram of the residuals will help you to determine whether the residuals are normally distributed. Plots of the residuals against fitted values and residuals against explanatory values will allow you to check whether the variance is constant. If it is not, you will see a funnel shape like the one shown in in Figure 11-4a. These plots also help you to spot patterns in the residuals, like the one shown in Figure 11-4b.

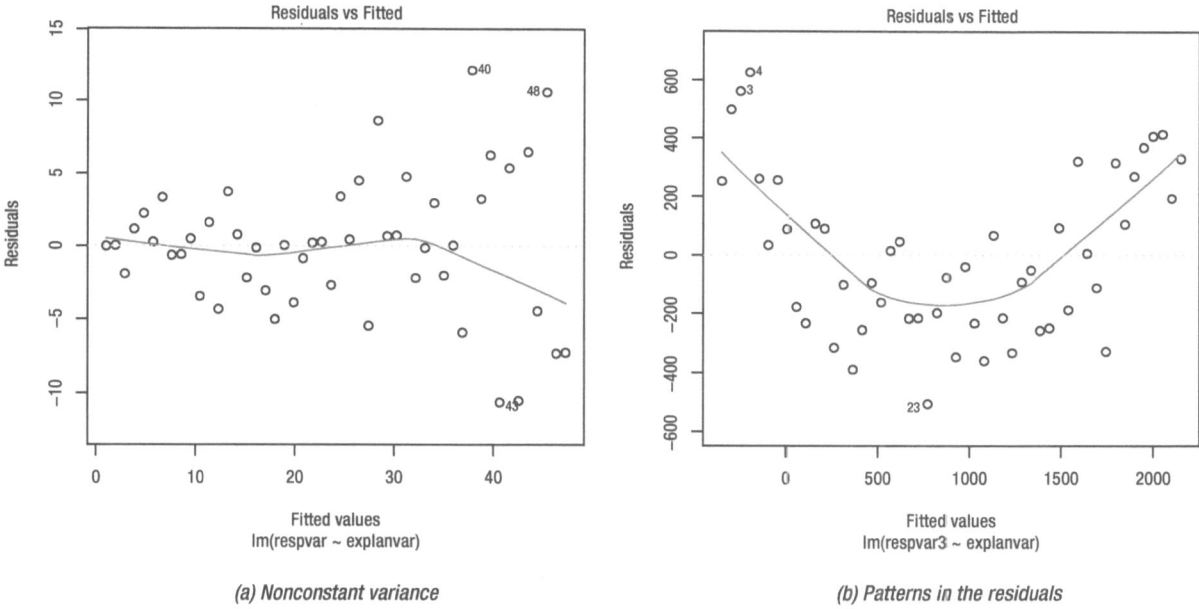

(a) Nonconstant variance   (b) Patterns in the residuals

Figure 11-4. *Residuals plots warning of issues with the model*

You can use the plot function to create residual plots for a model object:

```
> plot(lmobject, which=1)
```

Use the which argument to select from the following plots:

1. Residuals against fitted values

2. Normal probability plot of residuals

3. Scale-location plot

4. Residuals against leverage

I've omitted the numbers 4 and 6 which select plots of influence measures here, as you will learn about them in the next section.

Some other useful plots of residuals can be created:

- A histogram of residuals for checking the assumption of normality:

    ```
    > hist(dataset$resids)
    ```

- A plot of residuals against the response variable:

    ```
    > plot(resids~resp, dataset)
    ```

- A plot of the residuals against the explanatory variable:

    ```
    > plot(resids~var1, dataset)
    ```

EXAMPLE 11-13.
RESIDUAL PLOTS FOR THE TREEMODEL MODEL

Suppose that you wish to analyze the residuals for the treemodel model, to check that the assumptions of the model are met.

First, save the studentized residuals to the trees dataset:

```
> trees$Resids<-rstudent(treemodel)
```

Next, set up the graphics device to display six plots, as explained in Chapter 9 in the "Multiple Plots in the Plotting Area" section:

```
> par(mfrow=c(3,2))
```

Next, use the relevant commands to create the following plots: histogram of residuals; normal probability plot of the residuals; residuals against fitted values; residuals against response (Volume); residuals against explanatory variable (Girth); residuals against other variable of interest (Height):

```
> hist(trees$Resids)
> plot(treemodel, which=2)
> plot(treemodel, which=1)
> plot(Resids~Volume, trees)
> plot(Resids~Girth, trees)
> plot(Resids~Height, trees)
```

Figure 11-5 shows the result.

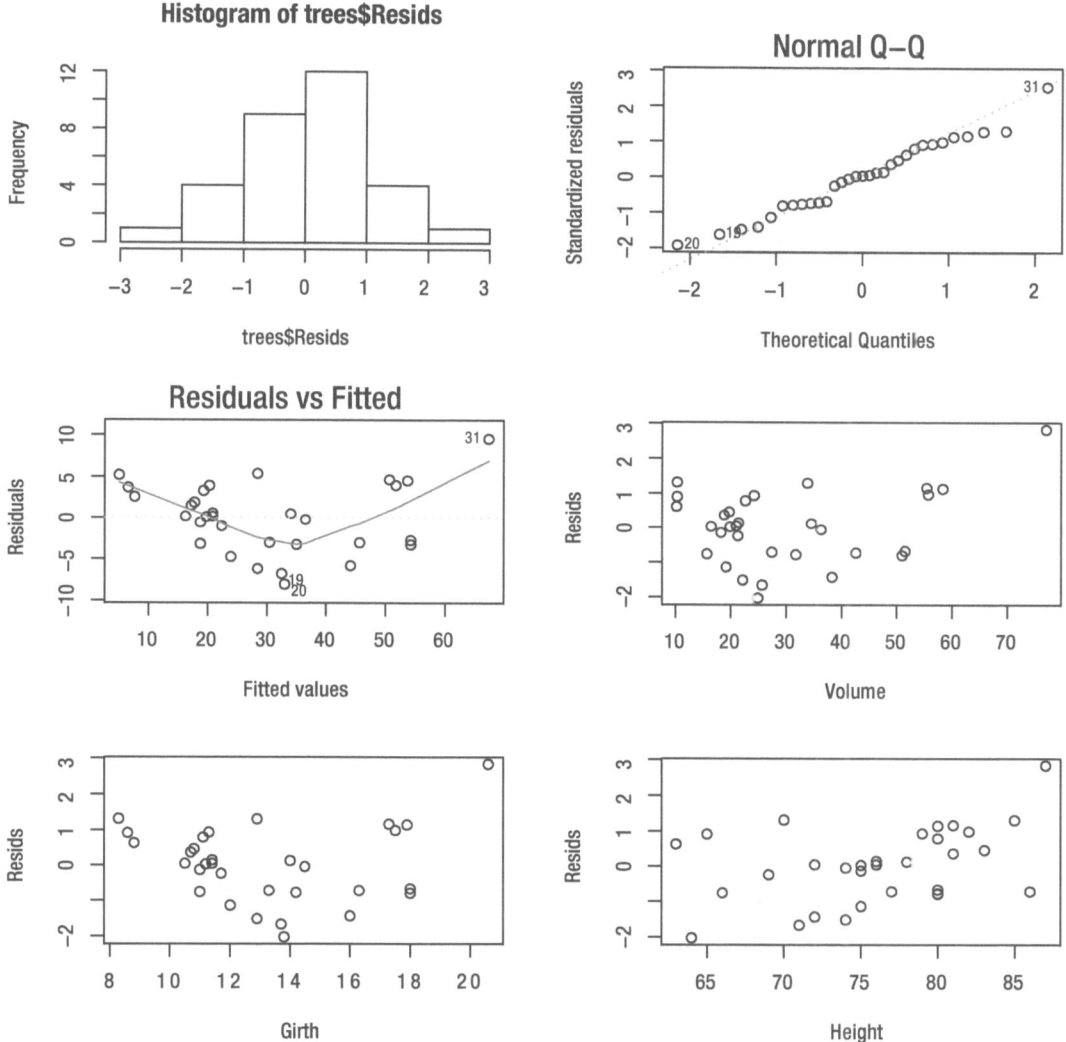

Figure 11-5. *Residual plots for the* treemodel *model*

From the histogram and normal probability plot, we can see that the residuals are approximately normally distributed.

In the plot of residuals against fitted values and the plot of residuals against girth, we can see that there is a slight pattern in the residuals. The residuals tend to be negative for trees with medium girth, and positive for trees with very small or very large girth. This suggests that adding polynomial terms to the model may improve the fit.

There are no obvious patterns in the plots of residual against volume and height. There are also no obvious outliers in any of the plots.

Leverage

The leverage helps to identify observations that have outlying values or unusual combinations for the explanatory variables. A large leverage value indicates that the observation may have a big influence on the model.

To calculate the leverage of each observation for a given model, use the hatvalues function:

```
> hatvalues(lmobject)
```

To create a plot of the residuals against the leverage, use the command:

```
> plot(lmobject, which=5)
```

To create a plot of the Cook's distances against the leverage, use the command:

```
> plot(lmobject, which=6)
```

Cook's Distances

The Cook's distance for an observation is a measure of how much the model parameters change if the observation is removed before estimation. Large values indicate that the observation has a big influence on the model.

To calculate the Cook's distances for a model, use the cooks.distance function:

```
> cooks.distance(lmobject)
```

You can create a plot of Cook's distance against observation number with the command:

```
> plot(lmobject, which=4)
```

and a plot of Cook's distance against leverage with the command:

```
> plot(lmobject, which=6)
```

EXAMPLE 11-14.
LEVERAGE AND COOK'S DISTANCES FOR THE TREEMODEL

Suppose that you want to create a set of plots for the treemodel regression model, to help identify any observations that may have a large influence on the model.

First set up the graphics device to hold four plots (see "Multiple Plots in the Plotting Area" in Chapter 9):

```
> par(mfrow=c(2,2))
```

Then create the three plots showing leverage and Cook's distances:

```
> plot(treemodel, which=4)
> plot(treemodel, which=5)
> plot(treemodel, which=6)
```

The result is shown in Figure 11-6.

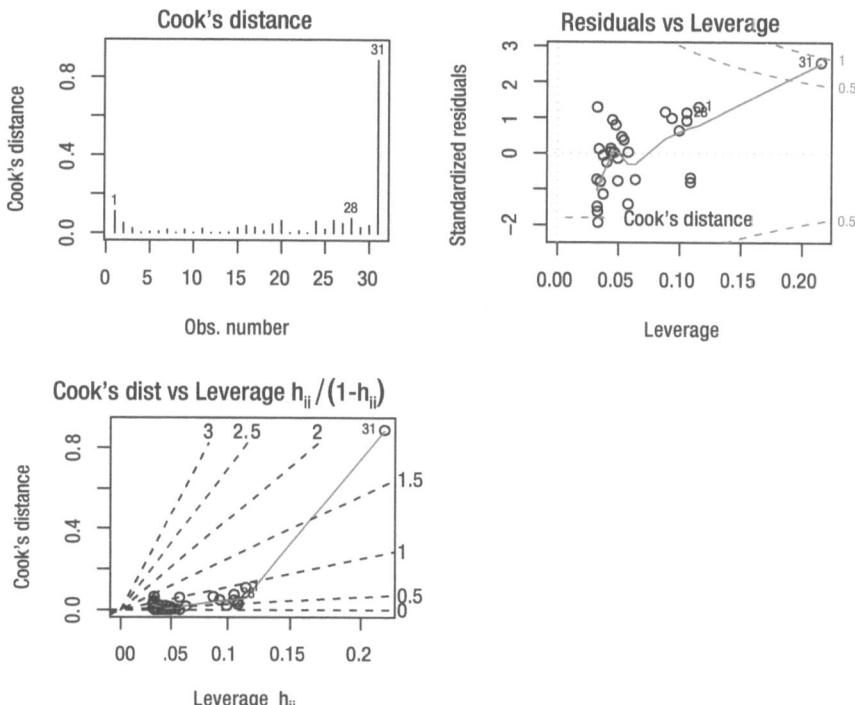

Figure 11-6. *Influence plots for* `treemodel` *model*

The contours on the second plot (residuals vs. leverage) divide areas with similar Cook's distances. The contours on the third plot (Cook's distance vs. leverage) divide areas with a similar standardized residual.

All three plots show that observation number 31 has a large Cook's distance and leverage, and also a fairly large standardized residual. This suggests that this observation has a big influence on the model and should be investigated further.

Making Predictions

Once you have built a model that you are happy with, you may want to use it to make predictions for new data. R has a convenient function for making predictions, called `predict`. To use this function, the new data must be arranged in a data frame (see "Entering Data Directly" in Chapter 2 for how to create data frames). The explanatory variables in the new dataset should be given identical names to those in the original dataset from which the model was built. It does not matter if the order of the variables is different, or if there are additional variables present.

Once your new data is arranged in a data frame, you can use the `predict` function:

```
> predict(lmobject, newdata)
```

The command creates a vector of predictions that correspond to the rows of the data frame. You can attach it to the data frame as a new variable:

```
> newdata$predictedvalues<-predict(lmobject, newdata)
```

You can also use the `predict` function to calculate confidence or prediction intervals for your predictions. Recall that confidence intervals only account for the uncertainty of the model estimation, while prediction interval also account for natural random variation in the response variable.

To calculate a confidence interval, set the `interval` argument to `"confidence"` and for a prediction interval, set it to `"prediction"`. You can adjust the size of the interval with the `level` argument:

```
> predict(lmobject, newdata, interval="confidence", level=0.99)
```

EXAMPLE 11-15.
MAKING PREDICTIONS USING TREEMODEL

Suppose that you want to use the `treemodel` regression model to estimate the volume of three trees with girths of 17.2, 12.0, and 11.4 inches.

First put the new data into a data frame:

```
> newtrees<-data.frame(Girth=c(17.2, 12.0, 11.4))
```

To make the predictions and add them to the `newtrees` data frame, use the command:

```
> newtrees$predictions<-predict(treemodel, newtrees, interval="prediction")
```

Then view the contents of the dataset:

```
> newtrees
```

	Girth	predictions.fit	predictions.lwr	predictions.upr
1	17.2	50.18927	41.13045	59.24809
2	12.0	23.84682	14.98883	32.70481
3	11.4	20.80730	11.92251	29.69210

From the output, you can see that for a tree with a girth of 17.2 inches, the predicted volume is 50.2 cubic feet with a prediction interval of 41.1 to 59.2.

Summary

You should now be able to build a general linear model including factor variables, polynomial terms, and interaction terms where appropriate, and interpret the model coefficients. You should be able to add or remove terms from an existing model and apply the stepwise model selection procedure. You should be able to assess how well the model fits the data and use model diagnostics to determine whether the model is valid. For simple linear regression and polynomial regression models, you should be able to represent the model graphically. Finally, you should also be able to use your model to make predictions about future observations.

This table summarizes the main commands covered.

Task	Command
Build simple linear regression model	`lm(`*resp~var1*`}, `*dataset*`)`
Build multiple regression model	`lm(`*resp~var1+var2+var3*`, `*dataset*`)`
Build model with interaction term	`lm(`*resp~var1+var2+var1*`:`*var2*`, `*dataset*`)`
Build model with interaction terms to a given order	`lm(`*resp~*`(`*var1+var2+var3*`)^2, `*dataset*`)`
Build factorial model	`lm(`*resp~var1*`*`*var2*`*`*var3*`, `*dataset*`)`
Build polynomial regression model	`lm(`*resp~var1*`+I(`*var1*`^2)+I(`*var1*`^3), `*dataset*`)`
Build model with log-transformed response variable	`lm(log(`*resp*`)~`*var1+var2+var3*`, `*dataset*`)`
Build model without intercept term	`lm(`*resp~*`-1+`*var1+var2+var3*`, `*dataset*`)`
Build null model	`lm(`*resp~*`1, `*dataset*`)`
Update a model	`update(`*lmobject*`, ~.+`*var4*`)`
Stepwise selection	`step(`*lmobject*`)`
Summarize a model	`summary(`*lmobject*`)`
Coefficient estimates	`coef(`*lmobject*`)`
Confidence interval for coefficient estimate	`confint(`*lmobject*`)`
Plot line of best fit	`plot(`*resp~var1*`, `*dataset*`)` `abline(coef(`*lmobject*`))`
Raw residuals	`residuals(`*lmobject*`)`
Standardized residuals	`rstandard(`*lmobject*`)`
Studentized residuals	`rstudent(`*lmobject*`)`
Fitted values	`fitted(`*lmobject*`)`
Residuals and influence plots	`plot(`*lmobject*`, which=1)`
Leverage	`hatvalues(`*lmobject*`)`
Cook's distances	`cooks.distance(`*lmobject*`)`
Predictions	`predict(`*lmobject*`, `*newdata*`)`

You have now covered everything you need to perform the most common types of statistical analysis. For the curious reader, Appendix A will explain how to access additional functionality with the use of add-on packages.

APPENDIX A

■ ■ ■

Add-On Packages

Over five thousand add-on packages are available for R, which serve a wide variety of purposes. An add-on package contains additional functions and sometimes objects such as example datasets. This appendix explains how to find a package that serves your purpose and install it.

Viewing a List of Available Add-on Packages

To view a list of available add-on packages, follow these instructions:

1. Go to the R project website at www.r-project.org

2. Follow the link to CRAN (on the left-hand side)

3. You will be taken to a list of sites that host the R installation files (mirror sites), as shown in Figure A-1. Select a site close to your location

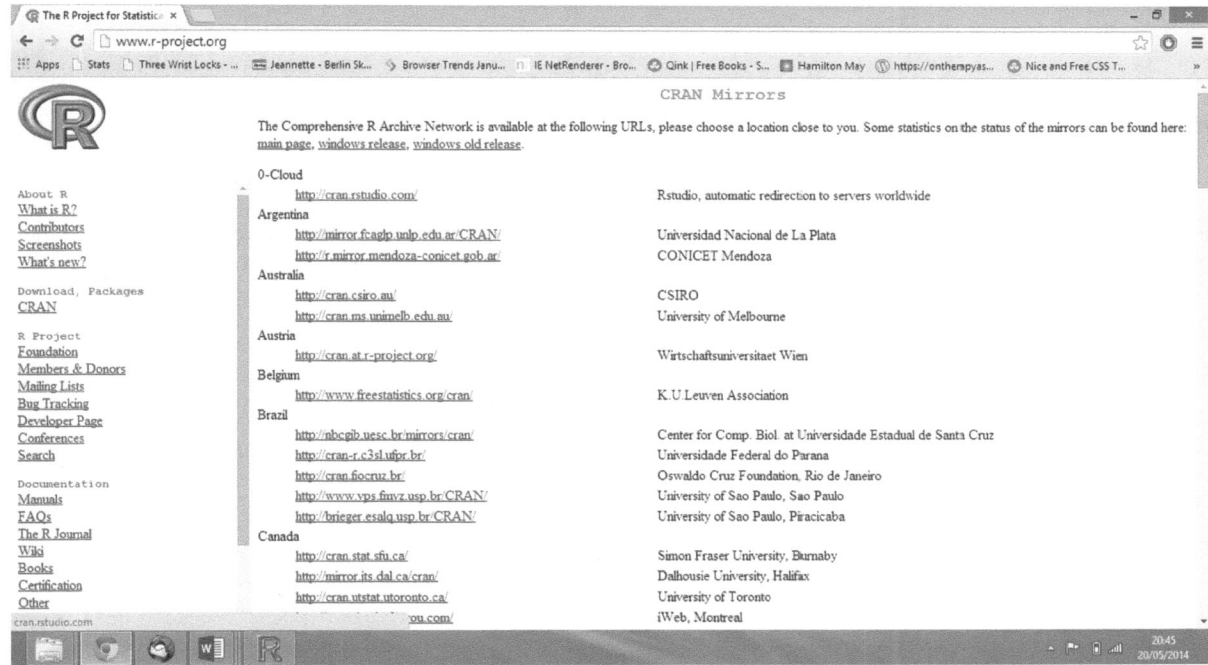

Figure A-1. *List of mirror sites*

4. Select Packages from the menu on the left-hand side

5. Select Table of available packages, sorted by name

A list of packages with descriptions of their purpose is displayed, as shown in Figure A-2. You can use the browser tools to search the list, usually by entering Ctrl+F or Cmd+F.

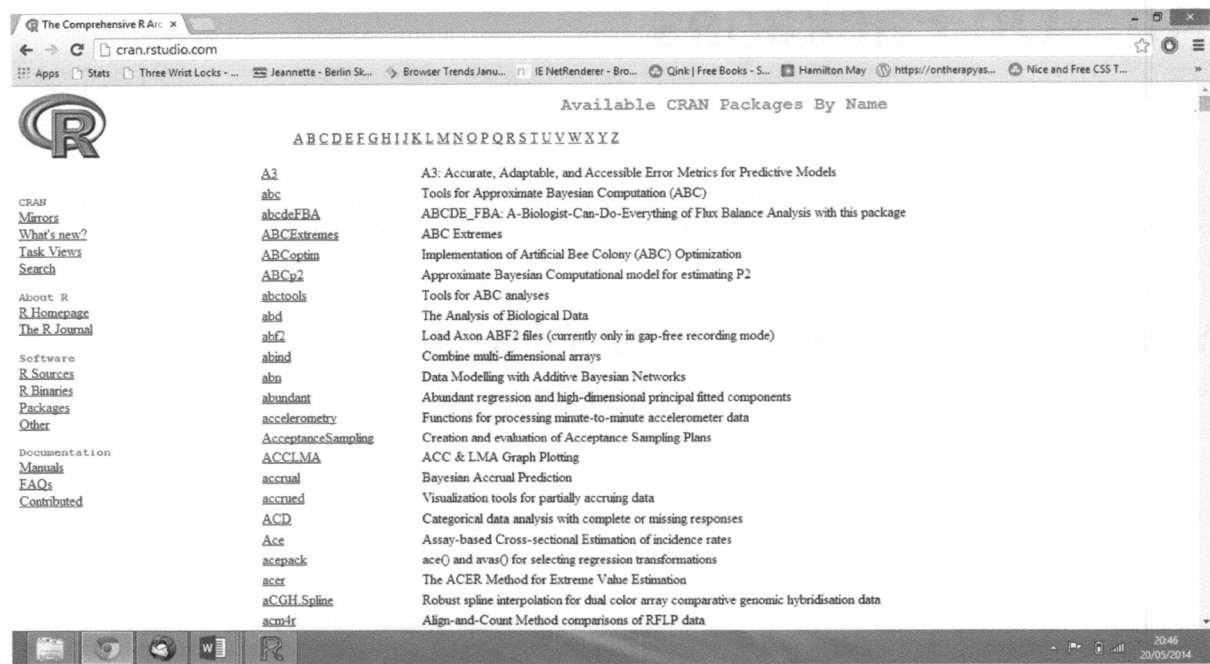

Figure A-2. *List of available package sorted by name*

On selecting a suitable package, you will see a package information page, as shown in Figure A-3. Here you will find a description of what the package does and a reference manual is available in pdf format. You will notice that the package is available to download, but you do not need to do this as it is simpler to install the package from within the R environment.

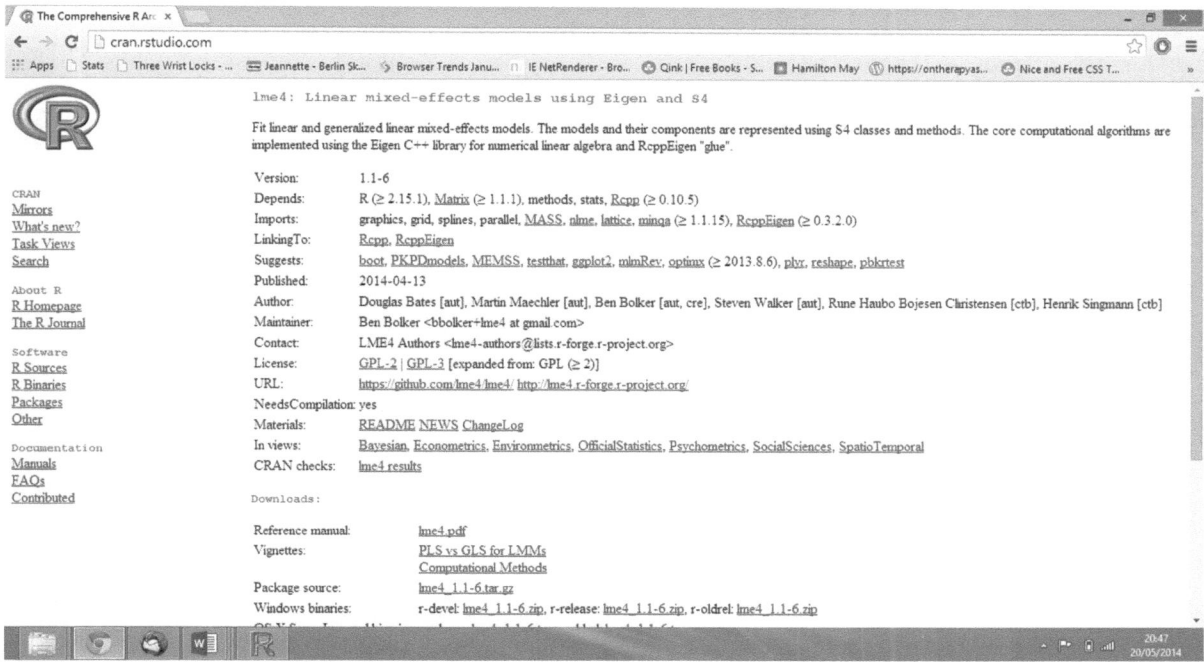

Figure A-3. *Package information page for the* lme4 *package*

Installing and Loading Add-On Packages

To use an add-on package, you must first install it, which only needs to be done once. There are a number of packages that are included with the R base installation (such as the foreign package that we used in Chapter 2), which do not need to be installed.

Once a package is installed, it must be loaded before you can use the functions within. The functions will be available for the remainder of the session, so you will need to load the package during each session that you intend to use it.

You can install and load packages from within the R environment, which is explained separately for Windows, Mac, and Linux users.

Windows Users

To install a package:

1. Select Install Package(s) from the Package menu

2. The first time you install a package, you will be prompted to select a mirror site, as shown in Figure A-4. Select a site close to your location

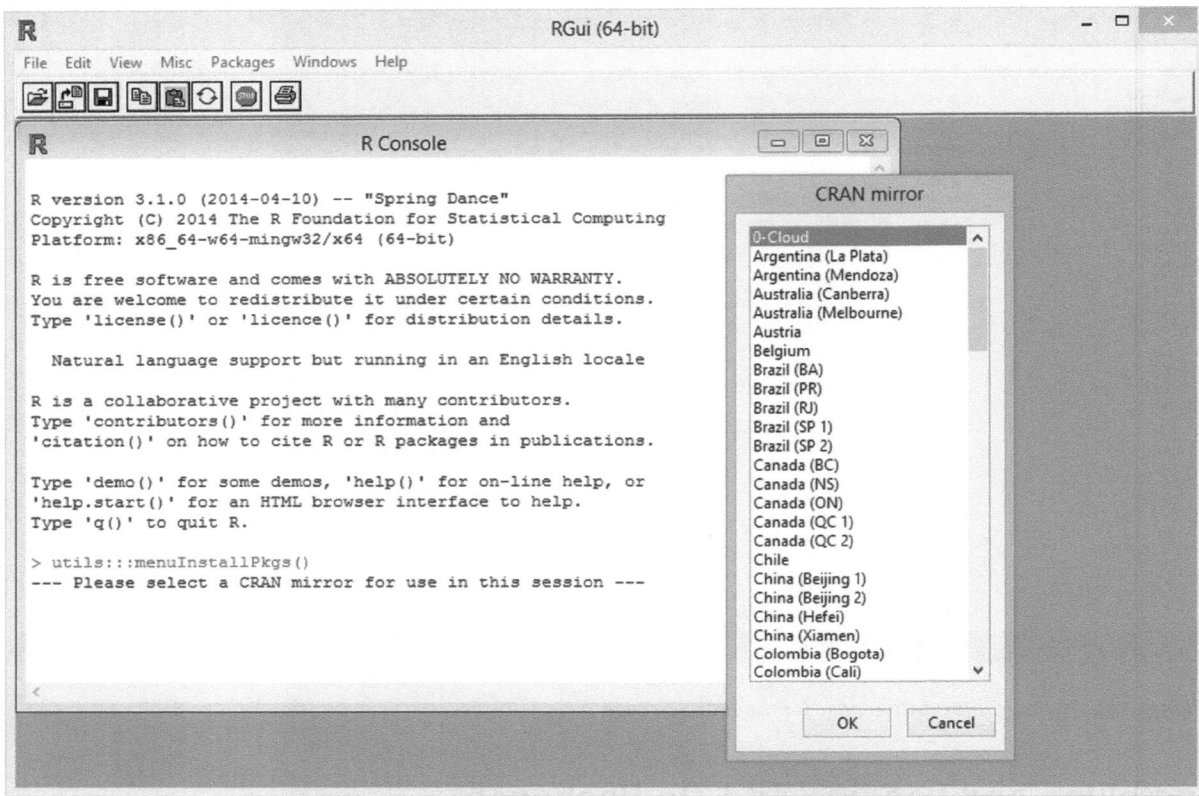

Figure A-4. *Selecting a mirror site (Windows)*

3. When prompted, selected the required package from the list, as shown in Figure A-5

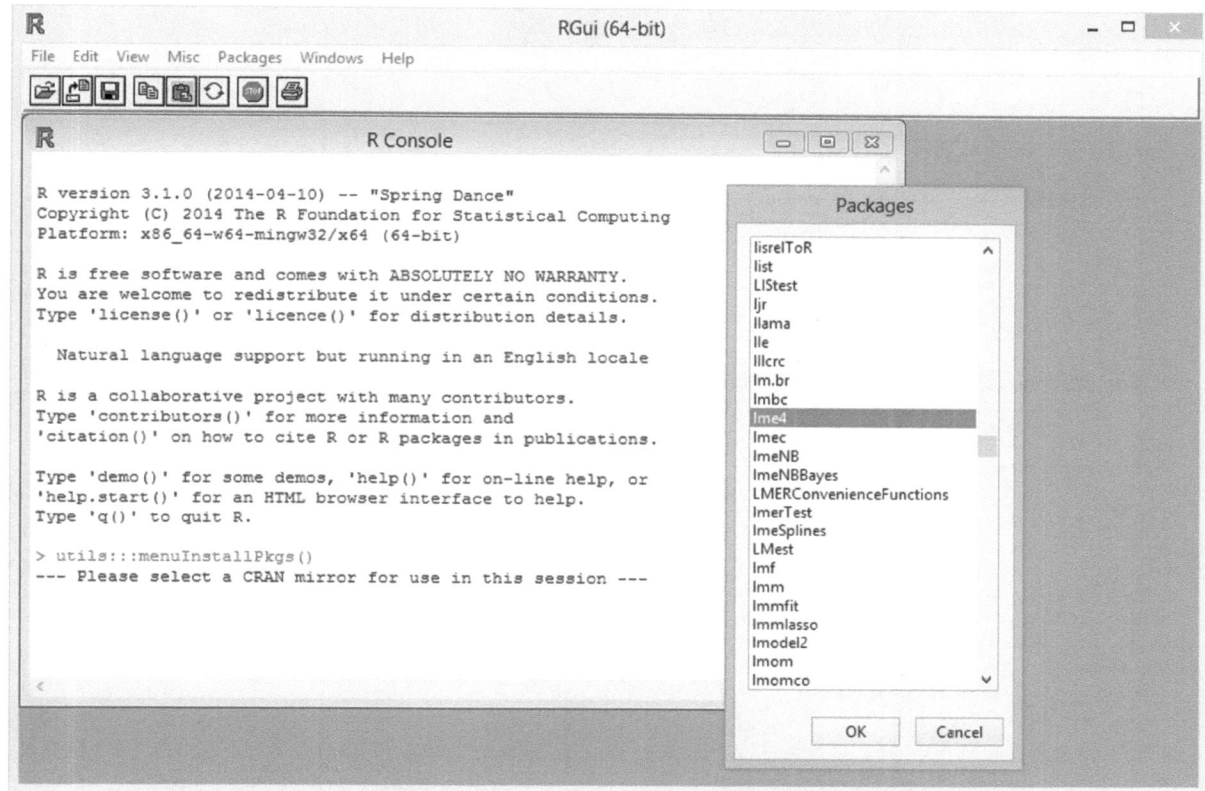

Figure A-5. *Selecting a package to install (Windows)*

To load a package:

1. Select Load Package from the Packages menu

2. When prompted, select the required package from the list

Mac Users

To install a package:

1. Select R Package Installer from the Packages & Data menu

2. When the R Package Installer appears, press Get List

3. A list of packages is displayed. Select the required package from the list, then select the Install Dependencies tick box and press Install Selected, as shown in Figure A-6

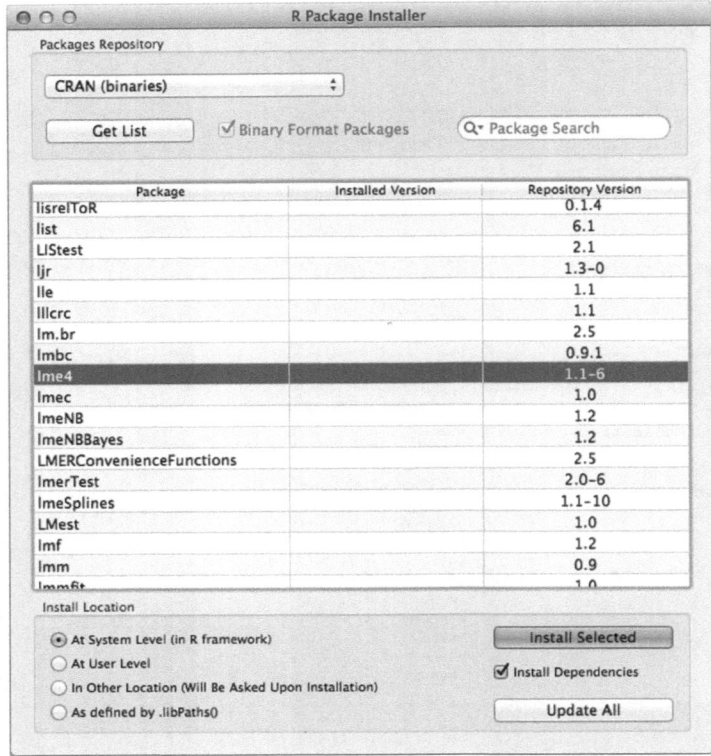

Figure A-6. *Selecting a package to install (Mac)*

4. Close the window

To load a package:

1. Select R Package Manager from the Packages & Data menu.

■ **Note** There is an issue in R version 3.1.0 for Mac which means that you may not be able to open the Package Manager. If you have this problem, you can download a patched version from `http://r.research.att.com/`. This should be resolved for versions 3.1.1 onwards.

---

2. Select the Status box next to the required package so that the status changes to Loaded, as shown in Figure A-7.

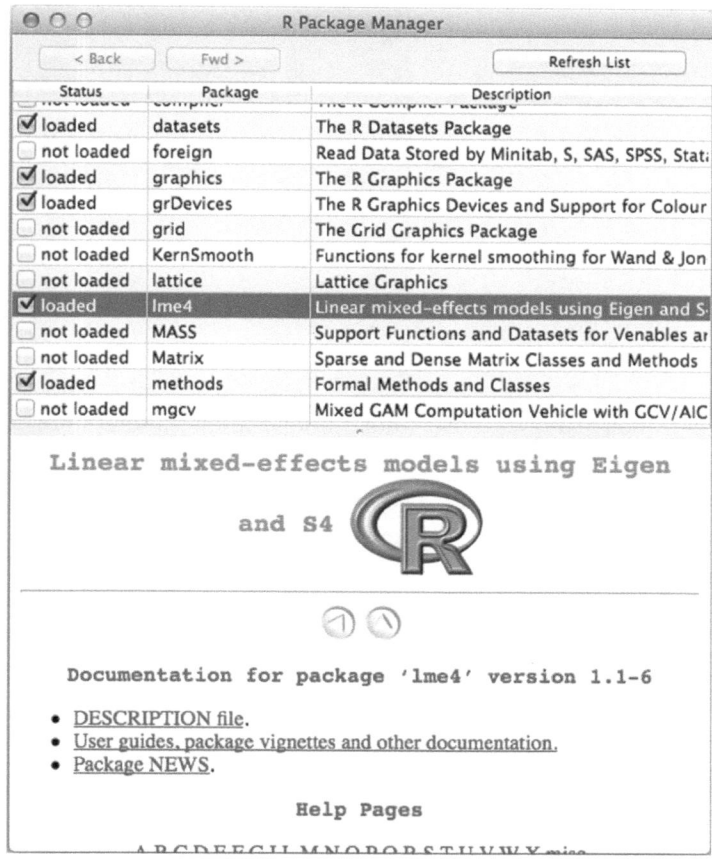

Figure A-7. *Loading a package (Mac)*

Linux Users

To install a package:

1. Enter the command:

   > download.packages(*packagename*, "*/home/Username/folder*").

   The file path gives the location in which to save the package

2. When prompted, select a mirror site close to your location

   To load a package, enter the command:

   > install.packages("*packagename*")

APPENDIX B

■ ■ ■

Basic Programming with R

As well being a statistical computing environment, R is also a programming language. The topic of R programming is beyond the scope of this book. However, this chapter gives a brief introduction to basic programming and writing your own functions.

When programming, use a script file so that you can edit your code easily. Remember to add plenty of comments with the hash symbol (#), so that your script can be understood by others or by yourself at a later date.

Creating New Functions

Throughout this book, you will have used many of R's built-in functions, but it is also possible to create your own. Creating your own function allows you to neatly bundle together a group of commands that perform a specific task that you want to use repeatedly.

To create a new function, the script takes the following general form:

```
functionname<-function(arg1name, arg2name, ..., argNname) {

    command(s)

    return(outputvalue)
}
```

All of the indentation and additional blank lines are optional, but they help to show the hierarchy of the program and are considered good programming practice.

The first line determines the name of the function and the names of the input arguments. You can include as many arguments as required, or none at all. Optionally, you can assign default values to the arguments:

```
functionname<-function(arg1name=value1, arg2name=value2, ..., argNname=valueN) {

    command(s)

    return(outputvalue)
}
```

This first line of the script ends with an opening curly bracket. After this is where the main content of the function begins. Generally, this is a set of commands that manipulate the input arguments in some way, or perform calculations from them, in order to create an output. The commands can make use of existing functions, including any that you have written yourself. Note that any objects that you create within a function exist only inside the function and are not saved to the workspace.

The `return` function determines the output of the function, which is either displayed in the console or assigned to an object whenever the function is used. It can either be a literal value such as a number or character string, or a it can be a vector, data frame or any other type of object.

Instead of using the `return` function in the final command, you can use the `print` function, which always displays the output in the console window even if the output is assigned to an object. Alternatively, you may want to create a function that produces a plot instead of giving an output value. In this case, the final command would be `plot` or another plotting function.

Finally, the script ends with a closing curly bracket.

Once you have written your function and run the script, the function is saved to the workspace as an object and is available to use for as long as the current workspace is loaded. You can use it just like a built-in function:

```
> functionname(value1, value2, ..., valueN)
```

EXAMPLE B-1.
A FUNCTION FOR FINDING THE CUBE ROOT OF A VALUE

This script creates a simple function, which takes a single value as input and calculates the cube root:

```
cube.root<-function(x) {

    y<-x^(1/3)
    return(y)

}
```

Try entering the function into a script file and running the program. Recall that to run a script file in the Windows environment, you must highlight the text that you want to run and then right-click and select Run line or selection, as shown in Figure B-1. Alternatively, you can highlight the text and press the Run button. Mac users should highlight the text and then press Cmd+Return.

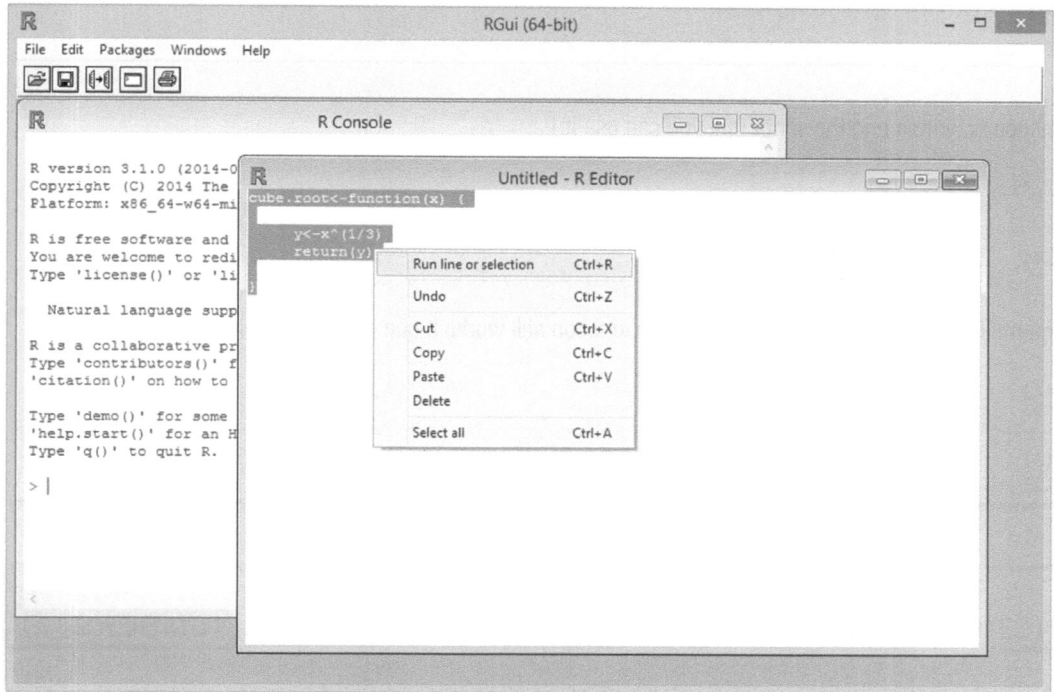

Figure B-1. *Entering a program into a script file and running (Windows)*

Once you have written the function and run the script, the `cube.root` function is available to use:

```
> cube.root(5)
```

```
[1] 1.709976
```

EXAMPLE B-2.
A FUNCTION FOR CALCULATING THE HYPOTENUSE OF A TRIANGLE

This script creates a simple function, which calculates the length of the hypotenuse of a triangle, given the lengths of the other two sides (recall that the formula is $c^2=a^2+b^2$):

```
hypot<-function(a=2, b=3) {

    c=sqrt(a^2+b^2)
```

```
    return(c)

}
```

Once the function is written and the script run, you can use it:

```
> hypot(6, 7)
```

```
[1] 9.219544
```

As the arguments have been given default values, the function still works if one or both of them are missing:

```
> hypot(b=2)
```

```
[1] 2.828427
```

EXAMPLE B-3.
A FUNCTION THAT CREATES A HISTOGRAM OF RANDOM NUMBERS

This script creates a function that generates a specified amount of random numbers from a standard normal distribution and plots them in a histogram:

```
randhist<-function(n) {

    vector1<-rnorm(n)      # Generate n random numbers
    # Plot random numbers in a histogram, add a title and remove the x-axis label:
    hist(vector1, main="Histogram of a random sample from a\nstandard normal distribution",
        xlab="")

}
```

Once the function is written and the script has been run, you can use it:

```
> randhist(50)
```

The result should be similar to Figure B-2.

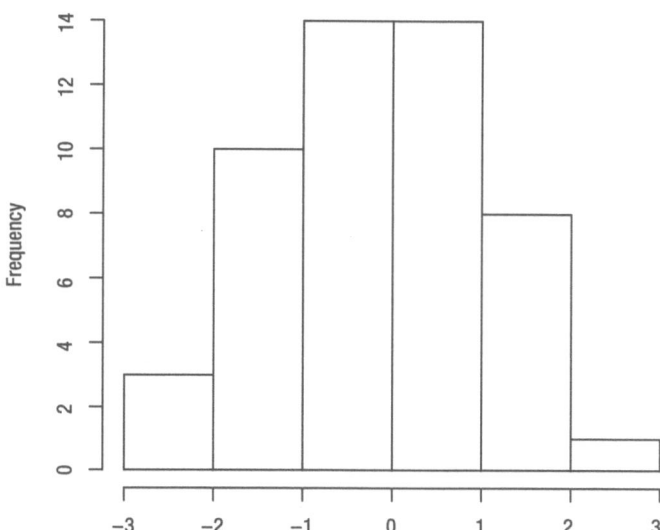

Figure B-2. *Histogram created by the* randhist *function*

Conditional Statements

Consider the cube.root function created in Example B-1. If you try to input a negative number, you get the result shown here.

```
> cube.root(-5)
```

```
[1] NaN
```

It would be nice if the function could return an error message to explain to the user what went wrong. In order to do this, the function would need to behave differently depending on whether the user inputs a positive number or a negative number. This is what a *conditional statement* allows you to do.

A conditional statement allows you to perform a command or set of commands only under certain circumstances (called the *condition*). Conditional statements add flexibility to your functions, as they allow the function to behave in different ways depending on the input.

There are a few types of conditional statement that allow you to do slightly different things. Before looking at these, you need to understand conditions and how they are constructed in R.

Conditions

In programming, a condition is an expression that can be either true or false. For example, the condition '4<5' (4 is less than 5) is true, whereas the condition '5==8' (5 is equal to 8) is false.

If you enter a condition at the command prompt, R tells you whether it is true or false:

```
> 6>8
```

```
[1] FALSE
```

A condition must contain at least one *comparison operator* (also known as a *relational operator*). In the preceding example, the comparison operator is > (greater than). Table B-1 gives a list of comparison operators that you can use to form conditions.

Table B-1. *Comparison operators*

Operator	Meaning
==	Equal to
<	Less than
<=	Less than or equal to
>	Greater than
>=	Greater than or equal to
%in%	In

The %in% operator compares a single value with all members of a vector:

```
> vector1<-c(4,2,1,6)
> 2 %in% vector1
```

```
[1] TRUE
```

Conditions containing only constant values are not very useful because we already know in advance whether they are true or false. More useful are conditions that include objects, such as 'object1<5' (the value of object1 is less than 5). Whether or not this condition is true depends on the value of object1:

```
> object1<-4
> object1<5
```

```
[1] TRUE
```

You can join two or more conditions to form a larger one, using the OR and AND operators.

The AND operator is denoted &. When two expressions are joined with the AND operator, both must be true in order for the whole condition to be true. For example, this condition is false because only one of the expressions is true:

```
> 3<5 & 7<5
```

```
[1] FALSE
```

This statement is true because both of the expressions are true:

```
> 3<5 & 7>5
```

```
[1] TRUE
```

The OR operator is denoted |. When two expressions are joined with the OR operator, the overall condition is true if either one or both of the expressions are true. For example, this condition is true because one of the expressions is true:

```
> 3<5 | 7<5
```

```
[1] TRUE
```

The condition is also true when both expressions are true:

```
> 3<5 | 7>5
```

```
[1] TRUE
```

You can negate a condition with the ! operator. This reverses the result of the condition:

```
> !3<5
```

```
[1] FALSE
```

If the condition is complex, you can use brackets to negate the entire condition:

```
> !(3<5 & 7<5)
```

```
[1] TRUE
```

If Statement

The simplest form of conditional statement is the *if* statement. The if statement consists of a condition and a command. When R runs an if statement, it first checks whether the condition is true or false. If the condition is true, it runs the command and if it is false it does not. The general form for the statement is shown here:

```
if (condition) command
```

You can also include a group of several commands in an if statement, by placing them between curly brackets:

```
if (condition) {

    commands to be performed if condition is true

}
```

The if statement is very useful as part of a function, as illustrated in the following examples.

EXAMPLE B-4.
FUNCTION FOR CALCULATING THE CUBE ROOT OF A VALUE (UPDATED)

This script creates an updated version of the cube.root function from Example B-1, which returns a warning message if the user inputs a negative number. Notice that it uses the warning function, which prints warning messages. R also has a function called stop, which causes the function to abort and prints an error message:

```
cube.root<-function(x) {

    y<-x^(1/3)                    # Calculate the cube root
    # If user enters a negative number, print warning message:
    if (x<0) warning("Cannot calculate cube root of negative number")
    return(y)

}
```

The updated function now returns a warning message only if the user input a negative number:

```
> cube.root(5)
```

```
[1] 1.709976
```

```
> cube.root(-5)
```

```
[1] NaN
Warning message:
In cube.root(-5) : Cannot calculate cube root of negative number
```

EXAMPLE B-5.
A FUNCTION FOR CALCULATING BODY MASS INDEX

A person's body mass index (BMI) is calculated from his or her height in meters and weight in kilograms using the formula:

$BMI = Weight/Height^2$

If imperial measurements are used (height in inches and weight in pounds), the formula is:

$BMI = Weight \times 702/Height^2$

This script creates a function to calculate the BMI from a height and weight. By default, the function calculates the BMI, assuming that the metric measurements have been supplied. If the user sets the `units` argument to `"imperial"`, the function makes the appropriate adjustment for imperial measurements. Notice the use of the if statement (shown in bold) to control whether the adjustment is made:

```
bmi<-function(height, weight, units="metric") {

    bmi<-weight/height^2              # Calculate BMI
    if (units=="imperial") bmi<-bmi*702   # Adjust for imperial measurements
    return(bmi)

}
```

Once the script has been run, you can use the function to calculate BMI using metric measurements:

```
> bmi(1.7, 70)
```

```
[1] 24.22145
```

or imperial measurements:

```
> bmi(66, 125, "imperial")
```

```
[1] 20.17332
```

If/else Statement

The *if/else* statement extends the if statement to include a second command (or set of commands) to be performed if the condition is false. The general form is shown here:

```
if (condition) command1 else command2
```

You can include groups of commands between curly brackets:

```
if (condition) {

    commands to be performed if condition is true

} else {

    commands to be performed if condition is false

}
```

You can also extend the if/else statement to accommodate three or more possible outcomes:

```
if (condition1) command1 else if (condition2) command2 else command3
```

When running the statement, R begins by checking whether the first condition is true or false. If it is true, then the first command is run. If the first condition is false, then R proceeds to check the second condition. If the second condition is true then the second command is run. Otherwise, the final command is run.

EXAMPLE B-6.
A FUNCTION FOR CLASSIFYING DATES

This script creates a function for classifying a date (given in the format ddmmmyyyy) as a weekend or a weekday:

```
day.type<-function(date) {

    date1<-as.Date(date, "%d%b%Y")      # Converts the date to date format
    # Determines whether date is a weekend or weekday and prints the
    # result:
    if (weekdays(date1) %in% c("Saturday", "Sunday")) return ("Weekend") else
        return("Weekday")

}
```

The function gives a different output depending on whether the input date is a weekend or a weekday:

```
> day.type("27JUN2014")
```

```
[1] "Weekday"
```

```
> day.type("28JUN2014")
```

```
[1] "Weekend"
```

EXAMPLE B-7.
A FUNCTION FOR CLASSIFYING HEIGHTS

This script creates a function that takes a height in centimeters as input, and gives a height category as output. Heights below 140 cm are classified as 'Short', heights between 140 cm and 180 cm as 'Medium', and heights over 180 cm as 'Tall'.

```
heightcat<-function(height) {

    if (height<140) return("Short") else if (height<180) return("Medium")
        else return("Tall")

}
```

Once you have run the script, you can use the function:

```
> heightcat(136)
```

```
[1] "Short"
```

```
> heightcat(187)
```

```
[1] "Tall"
```

The switch Function

In some circumstances, you can use the switch function as a compact alternative to using if/else statements with many possible outcomes. The switch function selects between a list of alternative commands, each of which must return a single value. R compares the input with a list of options, and if it finds a match then it performs the corresponding command. The final command (which is optional) is performed if the input does not match any of the options:

```
> switch(input, option1=command1, option2=command2, option3=command3, command4)
```

There is another use of the switch function which takes an integer value as input, and outputs the corresponding value from a list of values:

```
> switch(input, value1, value2, value3, value4)
```

EXAMPLE B-8.
FUNCTION TO PERFORM A SELECTED CALCULATION WITH TWO NUMBERS

This script creates a function that allows the user to give two numbers and a calculation type. The switch function is used to select from several commands, depending on which option the user selects:

```
calculator<-function(number1, number2, calctype="add") {

    result<-switch(calctype,
        "multiply"=number1*number2,
        "divide"=number1/number2,
        "add"=number1+number2,
        "subtract"=number1-number2,
        "exponent"=number1^number2,
        "Invalid calculation type"     # To cover all other possibilities
    )

    return(result)
}
```

Once the script has been run, you can use the function:

```
> calculator(2, 3, "divide")
```

```
[1] 0.6666667
```

```
> calculator(2, 3, "mean")
```

```
[1] "Invalid calculation type"
```

EXAMPLE B-9.
FUNCTION FOR GIVING THE NAME OF THE DAY OF THE WEEK

This script creates a function named week.day that takes a number from 1 to 7 as input, and returns a character string giving the corresponding day of the week:

```
week.day<-function(daynum) {

    dayname<-switch(daynum,
        "Monday",
        "Tuesday",
        "Wednesday",
        "Thursday",
        "Friday",
        "Saturday",
        "Sunday"
    )

    return(dayname)

}
```

Once you have run the script, you can use it:

```
> week.day(2)
```

```
[1] "Tuesday"
```

Loops

Consider this script, which creates a function that takes a single number as input, multiplies it by each of the numbers 1 to 10, and displays the results in the console window:

```
times.table<-function(x) {

    result<-x*1                               # Calculate result for x*1
    text<-paste(x, "times 1 equals", result)  # Create character string
                                              # giving the result
    print(text)                               # Print the result

    # Repeat for numbers 2 to 10:
    Result<-x*2
    text<-paste(x, "times 2 equals", result)
    print(text)

    result<-x*3
    text<-paste(x, "times 3 equals", result)
    print(text)

    # Continues to x*10

}
```

The function repeats the same three commands 10 times. It would be much more efficient if you could write the commands once and tell R to repeat them for each of the numbers 1 to 10. This is where *loops* become useful.

Loops allow you to repeat a command (or set of commands) a number of times. The two most important types of loop are the *for* loop and the *while* loop.

For Loop

The for loop allows you to repeat a command or set of commands a prespecified number of times. The for loop takes the following general form:

```
for (i in startvalue:endvalue) command
```

You do not have to use i for the repetition number. Any valid object name can be used; however, i is conventional. The startvalue and endvalue can be either constant values or object names. R will repeat the command for each of the values from startvalue to endvalue.

Alternatively you can give a nonsequential vector as shown here, or you can even give a vector of character strings. R will repeat the command once for each of the values in the vector.

```
for i in c(5,7,8,2,11) command
```

You can also include a set of commands to be repeated by enclosing them within curly brackets:

```
for (i in startvalue:endvalue) {

    commands to be repeated

}
```

EXAMPLE B-10.
FUNCTION FOR CALCULATING TIMES TABLES

This script creates a function that takes a single number as input, multiplies it by each of the numbers 1 to 10, and displays the results in the console window. The for loop (shown in bold) is used to repeat the calculation for each of the numbers 1 to 10:

```
times.table<-function(x) {

    for (i in 1:10) {
        result<-x*i                                # Calculate result
        # Create character string giving the result:
        text<-paste(x, "times", i, "equals", result)
        print(text)                                # Print the result
    }
}
```

Once the script has been run, you can use the function:

```
> times.table(5)
```

```
[1] "5 times 1 equals 5"
[1] "5 times 2 equals 10"
[1] "5 times 3 equals 15"
[1] "5 times 4 equals 20"
[1] "5 times 5 equals 25"
[1] "5 times 6 equals 30"
[1] "5 times 7 equals 35"
[1] "5 times 8 equals 40"
[1] "5 times 9 equals 45"
[1] "5 times 10 equals 50"
```

While Loop

The *while* loop is suitable when you want to repeat a command or set of commands until a given condition is satisfied, and you don't know in advance how many repetitions will be required in order to achieve this. The general form for the while loop is shown here:

```
while (condition) command
```

To include a group of commands, use curly brackets:

```
while (condition) {

    commands to be repeated

}
```

The commands within the loop should do something that will affect whether or not the condition is true. For example, this script will keep printing the value of i until it reaches 10, at which point it will exit the loop:

```
i<-1
while (i<10) {

    print(i)
    i<-i+1

}
```

If the commands within the loop do not affect whether the condition is true, R will keep processing the commands infinitely. Consider this example:

```
b<-2
while (b<3) {
    print(b)
    a<-4
}
```

As the command within the loop is unrelated to the condition, the condition continues to be true each time the loop is repeated. This causes R to keep repeating the loop indefinitely and to stop responding. If you find R is stuck repeating a loop, press the Esc key to cancel the commands.

EXAMPLE B-11.
FUNCTION FOR SIMULATING DIE ROLLS

The following script creates a function that simulates die rolls. The function keeps rolling imaginary dice until a six is rolled:

```
die.rolls<-function() {

    roll<-0                   # Create the object before using it in the loop

    while (roll!=6) {
        roll<-sample(1:6, 1) # Generate a random number between 1 and 6
        print(roll)
    }
}
```

Once the script is run, you can use the function:

```
> die.rolls()
```

```
[1] 1
[1] 1
[1] 5
[1] 6
```

```
> die.rolls()
```

```
[1] 1
[1] 3
[1] 3
[1] 1
[1] 3
[1] 6
```

Summary

You should now understand the basics of R programming and be able to create simple functions to perform routine tasks. You should be able to make your programs flexible by using appropriate statements to perform conditional execution, and use loops to perform sets of commands repeatedly.

This table summarizes the main statements covered.

Task	General form

Create a function	*functionname*<-function(*arguments*) {
	        command(s)
	        return(*output*)
	}

if statement	if (*condition*) *command*
if/else statement	if (*condition*) *command1* else *command2*
if/else statement (extended)	if (*condition1*) *command1* else if (*condition2*) *command2* else *command3*
switch function	switch(*input*, *value1*, *value2*, *valueN*)
	switch(*input*, option1=*command1*, option2=*command2*, optionN=*commandN*)
for loop	for (*i* in *startval:endval*) *command*
while loop	while (*condition*) *command*

Datasets

This appendix gives details and sources where relevant for each of the example datasets used in this book. The datasets are available as an R workspace file or as separate csv files with the downloads for this book (www.apress.com/9781484201404).

apartments

Description	Details of 32 one-bedroom apartments advertised for rent within a five-mile radius of Bishops Stortford, Hertfordshire, UK, in October 2012.	
Variables	Town	Location of apartment
	Furnished	Provided furnished (Yes or No)
	Price.Cat	Rental price category (per calendar month)
Source	www.rightmove.co.uk	
Used in	Chapter 6	

bigcats

Description	Average weights for four big cat species	
Variables	Name	Name of species. Note that first instant of Leopard refers to *Pantera Pardus* and the second instance refers to *Unica unica* (Snow Leopard)
	Weight	Mean weight in kilograms for male of species
Source	Url: dialspace.dial.pipex.com/agarman/facts1.htm	
Used in	Chapter 4	

bottles

Description	Dataset giving the volume of liquid within 30 bottles of soft drink randomly selected from a production line. This is fictional data
Variables	Volume Volume (milliliters)
Used in	Chapters 5, 10

brains

Description	Brain volume for 10 pairs of monozygotic twins, measured using magnetic resonance imaging and computer-based image analysis techniques	
Variables	Pair	Pair identifier
	Twin1	Total brain volume of first-born twin (cubic centimeters)
	Twin2	Total brain volume of second-born twin (cubic centimeters)
Source	This data is taken from the article "Brain Size, Head Size, and IQ in Monozygotic Twins," by Tramo, M. J., et al. and published in Neurology 1998; 50:1246-1252. Reproduced with permission. Url: lib.stat.cmu.edu/datasets/IQ_Brain_Size	
Used in	Chapter 10	

CIAdata1, CIAdata2

Description	Demographic data for seven European countries	
Variables	country	Country name
	lifeExp	Life expectancy (years)
	urban	Living in urban areas (%)
	pcGDP	Per capita gross domestic product ($US)
Source	Collected from the CIA World Factbook on August 5, 2012 Url: https://www.cia.gov/library/publications/the-world-factbook/	
Used in	Chapters 4, 9	

coffeeshop

Description	Total sales at a coffee shop over a five-day period. This is fictional data	
Variables	Date	Date in format dd/mmm/yyyy
	Sales	Sales for the day (£)
Used in	Chapter 3	

concrete

Description	The results of an experiment to determine the best concrete mix	
Variables	Cement	Cement type (I or II)
	Additive	Additive (A or B)
	Additive.Dose	Additive dose (0.3%, 0.4%, or 0.5%)
	Density	Density (grams per cubic centimeter)
Source	The experiment was conducted in Santiago, Chile, in 2007	
Used in	Chapter 11	

CPIdata

Description	Consumer price index data for six countries (2012)	
Variables	country	Country name
	CPI	Consumer price index (relative to New York at 100)
Source	Url: www.numbeo.com/cost-of-living/rankings_by_country.jsp	
Used in	Chapter 4	

customers

Description	The names and addresses of five customers living in the area of Reading, Berkshire, UK. This is fictional data
Variables	Name Character string giving customer's full name
	Address Character string giving customer's full address
Used in	Chapter 3

endangered

Description	Conservation status of four big cat species
Variables	Name Name of species
	Status Conservation status
Source	Url: www.bigcats.com/redlist.php
Used in	Chapter 4

fiveyearreport

Description	UK Sales (including VAT) for the years 2007 to 2011 for the Tesco, Sainsburys, and Morrisons supermarket chains
Variables	Year Year (2007–2011)
	Tesco UK sales including VAT (£M) for Tesco
	Sainsburys UK sales including VAT (£M) for Sainsburys
	Morrisons UK sales including VAT (£M) for Morrisons
Source	Data collected from respective annual reports for 2007–2011 Url: www.tescoplc.com/files/pdf/reports/tesco_annual_report_2011.pdf www.tescoplc.com/files/pdf/reports/annual_report_2010.pdf (p16) www.tescoplc.com/files/pdf/reports/annual_report_2009.pdf (p34) www.tescoplc.com/files/pdf/reports/annual_report_2008.pdf (p5) www.tescoplc.com/files/pdf/reports/annual_report_2007.pdf (p3) www.j-sainsbury.co.uk/investor-centre/financial-performance/5-year-summary/ www.morrisons.co.uk/corporate/2011/annualreport/investor-information/five-year-summary-results/
Used in	Chapter 9

flights

Description	Flight data for seven flights departing from Southampton Airport on January 12 and 13, 2012
Variables	Date Date of flight in format dd/mm/yyyy
	Time Time of flight in format hh:mm
	Flight.Number Alphanumeric flight number
	Destination Name of destination city
Source	Url: www.southamptonairport.com
Used in	Chapter 3

fruit

Description	Dataset of UK fruit prices in August 2012
Variables	Product Product name
	Price Sale price (£)
	Unit Sale unit
Source	Url: www.sainsburys.co.uk
Used in	Chapter 3

grades1

Description	Fictional dataset giving the grades of 15 students belonging to three classes labeled A, B, and C.
Variables	ClassA Grades of students in class A (%)
	ClassB Grades of students in class B (%)
	ClassC Grades of students in class C (%)
Used in	Chapters 4, 10

people

Description	Physical characteristics for a sample of 16 people. Sample selected using nonrandom methods. Data is self-reported	
Variables	Subject	Respondent number
	Eye.Color	Eye color (Blue, Green, or Brown)
	Height	Height (centimeters)
	Hand.Span	Hand span of left hand (millimeters)
	Sex	Sex (1=Male, 2=Female)
	Handedness	Handedness (L=left-handed, R=right-handed)
Used in	Chapter 3	

people2

Description	This dataset is a clean version of the people dataset.	
Variables	Subject	Respondent number
	Eye.Color	Eye color (Blue, Green, or Brown)
	Height	Height (centimeters)
	Hand.Span	Hand span of left hand (millimeters)
	Sex	Sex (Male or Female)
	Handedness	Handedness (Left or Right)
	Height.Cat	Height category (Tall, Medium, or Short)
Used in	Chapters 6, 8, 11	

powerplant

Description	Thirty measurements of pressure, temperature, and output for a gas electrical turbine	
Variables	Pressure	Pressure inside turbine (millibars)
	Temp	Temperature inside turbine (degrees Celsius)
	Output	Output of turbine (megawatts)
Source	Collected from a gas electrical turbine at a UK power station in 2010	
Used in	Chapter 11	

pulserates

Description	Pulse data for four people. Sample selected using nonrandom methods. Data is self-reported	
Variables	Patient	Patient identifier
	Pulse1	First pulse reading (beats per minute)
	Pulse2	Second pulse reading (beats per minute)
	Pulse3	Third pulse reading (beats per minute)
Used in	Chapter 3	

resistance

Description	Gives the results of a simple experiment to compare the cubic resistance of four concrete formulations, at three, seven, and fourteen days after setting	
Variables	Formula	A (Huechuraba Aggregate + Additive A) B (Huechuraba Aggregate + Additive B) C (Mauro Aggregate + Additive A) D (Mauro Aggregate + Additive B)
	Day3	Cubic resistance (kilograms per square meter), measured three days after setting
	Day7	Cubic resistance (kilograms per square meter), measured seven days after setting
	Day14	Cubic resistance (kilograms per square meter), measured fourteen days after setting
Source	Experiment conducted in Santiago, Chile, in 2007	
Used in	Chapter 4	

supermarkets

Description	Data for four UK supermarket chains	
Variables	Chain	Name of chain
	Stores	Number of stores in the UK
	Sales.Area	Sales area (1,000 square feet)
	Market.Share	Market share (%)
Source	Data for number of stores and total sales area is collected from the respective 2011 annual reports: www.tescoplc.com/media/417/tesco_annual_report_2011_final.pdf www.j-sainsbury.co.uk/investor-centre/reports/2011/annual-report-and-financial-statements-2011/ www.morrisons.co.uk/Documents/Morrisons-Annual-Report-2011.pdf Market share is collected from Kantar Worldpanel: www.kamcity.com/namnews/mktshare/2011/kantar-march11.htm	
Used in	Chapter 2	

vitalsigns

Description	Measurements of systolic blood pressure, diastolic blood pressure, and pulse rate for four patients. This is fictional data	
Variables	subject	Patient identifier
	test	Name of parameter (SysBP, DiaBP, Pulse)
	result	Systolic blood pressure (mmHg), diastolic blood pressure (mmHg), or pulse (beats per minute)
Used in	Chapter 4	

WHOdata

Description	Data on alcohol consumption and mortality rate for five countries	
Variables	alcohol	Alcohol consumption per adult over 15 years (liters of pure alcohol per person per year)
	mortality	Adult mortality rate (probability of dying between 15 and 60 years, per 1000 of population)
Source	Collected from the WHO website: apps.who.int/ghodata	
Used in	Chapter 4	

Index

Get the eBook for only $10!

> Now you can take the weightless companion with you anywhere, anytime. Your purchase of this book entitles you to 3 electronic versions for only $10.

This Apress title will prove so indispensible that you'll want to carry it with you everywhere, which is why we are offering the eBook in 3 formats for only $10 if you have already purchased the print book.

Convenient and fully searchable, the PDF version enables you to easily find and copy code—or perform examples by quickly toggling between instructions and applications. The MOBI format is ideal for your Kindle, while the ePUB can be utilized on a variety of mobile devices.

Go to www.apress.com/promo/tendollars to purchase your companion eBook.

All Apress eBooks are subject to copyright. All rights are reserved by the Publisher, whether the whole or part of the material is concerned, specifically the rights of translation, reprinting, reuse of illustrations, recitation, broadcasting, reproduction on microfilms or in any other physical way, and transmission or information storage and retrieval, electronic adaptation, computer software, or by similar or dissimilar methodology now known or hereafter developed. Exempted from this legal reservation are brief excerpts in connection with reviews or scholarly analysis or material supplied specifically for the purpose of being entered and executed on a computer system, for exclusive use by the purchaser of the work. Duplication of this publication or parts thereof is permitted only under the provisions of the Copyright Law of the Publisher's location, in its current version, and permission for use must always be obtained from Springer. Permissions for use may be obtained through RightsLink at the Copyright Clearance Center. Violations are liable to prosecution under the respective Copyright Law.